TEMPLE GRANDIN'S
GUIDE TO
WORKING WITH FARM ANIMALS

TEMPLE GRANDIN'S
GUIDE TO
WORKING WITH
FARM ANIMALS

Safe, Humane Livestock Handling
Practices for the Small Farm

Storey Publishing

The mission of Storey Publishing is to serve our customers by
publishing practical information that encourages
personal independence in harmony with the environment.

EDITED BY Deborah Burns
ART DIRECTION AND BOOK DESIGN BY Carolyn Eckert
TEXT PRODUCTION BY Erin Dawson
INDEXED BY Nancy D. Wood

COVER PHOTOGRAPHY BY © Jason Houston, back and
front (all except top center); © Shawn Linehan, front
(top center) and spine
INTERIOR PHOTOGRAPHY BY © Jason Houston, ii–iii,
vi–vii, viii, 2–7, 10, 13 (all except top left), 15 (top &
bottom rows), 16 (bottom), 19, 20 (top), 22, 28 (middle
left), 33, 34, 39 (top right, middle left), 42, 44, 45, 54,
56, 60–61, 70, 79, 124–125, 135, 146, 169, 182–183;
© Shawn Linehan, v, 8, 11, 21, 62, 81, 90; © Temple
Grandin: i, 12, 16 (top), 17 (bottom), 32 (bottom), 55, 66,
67, 69, 73 (top), 97, 104, 106 (bottom), 107, 110, 136,
138, 160
For additional photography credits, please see page 181.

ILLUSTRATIONS BY © Elayne Sears
DIAGRAMS BY Ilona Sherratt, adapted from designs by
Temple Grandin and Mark Deesing, unless noted

Portions of the text in this book have been excerpted from
Dr. Grandin's previous book, *Humane Livestock Handling*
(Storey, 2008).

Storey Publishing
210 MASS MoCA Way
North Adams, MA 01247
storey.com

Printed in China by R.R. Donnelley
10 9 8 7 6 5 4 3 2 1

Library of Congress Cataloging-in-Publication Data

Names: Grandin, Temple, author.
Title: Temple Grandin's guide to working with farm animals /
by Temple Grandin.
Other titles: Guide to working with farm animals
Description: North Adams, MA : Storey Publishing, 2017. |
Includes bibliographical references and index.
Identifiers: LCCN 2016055528 (print) | LCCN 2016056915
(ebook) | ISBN 9781612127446 (pbk. : alk. paper) |
ISBN 9781612127606 (hardcover : alk. paper) | ISBN
9781612127453 (ebook)
Subjects: LCSH: Livestock. | Domestic animals.
Classification: LCC SF65.2 .G72 2017 (print) | LCC SF65.2
(ebook) | DDC 636—dc23
LC record available at https://lccn.loc.gov/2016055528

DEDICATED TO ALL THE PEOPLE WHO
HELPED ME START MY CAREER:
My mother, Eustacia Cutler
William Carlock, my high school science teacher
Ann Brecheen on her Arizona ranch
Ted Gilbert, Red River Cattle Co.
Gary Oden, McElhaney Cattle Co.
and Jim Uhl, Agate Construction

CONTENTS

Introduction

MY PREVIOUS BOOK, *Humane Livestock Handling,* had many methods, recommendations, and handling-facility designs suitable for large livestock operations. This book is aimed at smaller producers and at people getting started with farm animals. Less-elaborate handling facilities are required for a small number of tame animals that have daily interaction with their caretakers. Here you will find handling system designs for smaller farms ranging from 3 to 35 cattle or 5 to 100 sheep and goats. (Since cattle, sheep, and goats are herd animals, it is advisable to have more than one of each.) All the behavior information has been updated with information from both practical experience and the latest research on animal behavior.

Grazing animals such as cattle, sheep, horses, goats, pigs, bison, and llamas play an important role in societies worldwide. Providing humane treatment throughout their lives is a responsibility that producers should willingly embrace. The results are both gratifying and profitable. Raising animals kindly and caring for them in a way that makes them content ensures good health and allows them to grow faster.

Calm animals are easier to handle than frightened, agitated animals. If cattle or other animals become agitated, a 20- to 30-minute rest is required to calm them down. Numerous research studies clearly show that animals that remain calm during handling have increased weight gain, better reproduction, and fewer bruises or injuries.

Stress is punishing on the body. Keep animals as calm as possible for veterinary procedures to help them quickly overcome the stress of restraint. Livestock producers have learned that when rough, stressful handling practices are eliminated, cattle resume eating a full day earlier.

Make Time for Simple Observation

One of the most valuable things you can do as a stockperson is schedule time every day simply to observe your animals — how they spend their time, how they move and forage and feed, how they interact with one another, how they respond to you. This is the best way to understand their nature, learn the dynamics of the herd, and know what is normal and what is not. And by being a good observer, you can further improve your animal-handling technique and truly learn the art of stockmanship.

Choosing Which Type of Livestock to Raise

Before starting out with livestock, you must decide which sector of the industry you will work in. This decision will determine your work commitment, the type of handling facilities you will need, and your marketing plan.

DAIRY, MEAT, OR FIBER?

Will it be dairy or meat/wool? Dairying requires a greater labor commitment than rearing animals for meat or wool, since goats and cows need a tight milking schedule with no flexibility. (The only exception to this is using automated robotic milking equipment — very expensive for an individual but a viable option for an established dairy.) Typical milking

Who Are the Grazing Animals?

Grazing animals are all of the hoofed prey species that are able to subsist on grasses, scrub, browse, and other plants. Common grazing animals — often raised by humans for meat, milk, fiber, recreation, or work — are cattle, sheep, goats, pigs, deer, bison, elk, llamas, alpacas, horses, and donkeys.

facilities range from a simple stanchion, for a few goats or a single cow, to an elaborate milking parlor for a large flock or herd.

Raising animals for meat or wool allows for a more flexible work schedule, and certain breeds are more labor-intensive than others. Sheep breeds are available in two types, wool sheep and hair sheep. Hair sheep do not have to be sheared. Some people call them "easy-care sheep." This reduces labor for somebody primarily interested in producing meat.

OTHER WAYS GRAZING ANIMALS CAN EARN THEIR KEEP

Some innovative producers have developed a side business of grazing their flocks in areas that need weed control. These areas can vary from subdivisions to vineyards to vacant lots. The sheep are fenced in with electric netting, and cattle trained to electric fences can often be confined with a single electric wire. Sheep are excellent at digging deep and destroying invasive vegetation.

Another highly sustainable use of livestock is grazing stubble after a crop has been harvested. This can form a sustainable, mutually beneficial relationship between the livestock producers and the

Above, a sheep, two goats, and a llama all look at the person entering their pen. Animals want to see you.

farmer. The animals fertilize the field and they get to eat economical forage. When done correctly, the soil is improved. Grazing animals can also be used as part of a sustainable program of crop rotation.

MARKETS

If you plan to raise livestock for meat, there are several different markets for your products. For very small producers, there are farmer's markets, ethnic markets where goat and sheep meat is part of the culture, direct marketing to individual customers, and the local auction. Larger producers can market directly to restaurants and to upscale retailers who promote local products — and you can raise livestock to fit the specifications of those retailers.

Another specialized market is fine wool for spinners. Clean fine wool can sell for high prices, and you can market it over the Internet. Raising fine wool

will demand more work than with raising meat sheep, however, because greater care is required to keep the fleece clean. Many producers put jackets on their fine wool animals. To place jackets on sheep in a low-stress manner will require that you spend time with sheep to gentle them. Fine wool sheep will need to be housed where they can be kept out of mud.

Consumers Value Good Handling

Learning good stockmanship will help you become a successful livestock producer and improve animal welfare. Consumers are increasingly concerned about where their food comes from and how animals are raised. They want to be assured that you really care about your animals and that you provide for their welfare.

Reducing illness is possible when animals are calm in their environment and handled humanely. When animals get sick, they end up having to be separated; besides being off feed and losing valuable weight, they need expensive medications. Producers who raise natural or organic beef, pork, lamb, and other meats know that keeping animals healthy is essential because sick animals that have been treated with antibiotics cannot be sold to the U.S. organic market. Injection-site damage compromises meat quality and, even worse, if the animal has prolonged sickness, the damage is permanent. A lower-grade meat is almost guaranteed because marbling is reduced.

For all species of grazing animals, one goal of low-stress handling is to move all animals at a walk or trot. Cattle and bison can injure themselves when they become agitated. When they go running through a working facility, they crash into fences, gates, and other animals and risk falling or injuring themselves.

Every bruise directly affects meat quality. Old bruises cause localized areas of tough meat. Fresh bruises at the meat plant cause huge losses because the bruised meat must be cut out and discarded. There are many good cuts of meat on the shoulder and brisket area, and when cattle charge into a head gate, that's the area getting hit. The muscles can be severely bruised even though the hide is undamaged. And even if a bruise has healed by the time the animal goes to market, there can still be an area of tough meat.

The bottom line: calm handling is supremely important for both animal and human welfare, and for the economics of your farm.

Grazing animals should always be moved at a slow, relaxed pace, ideally a walk.

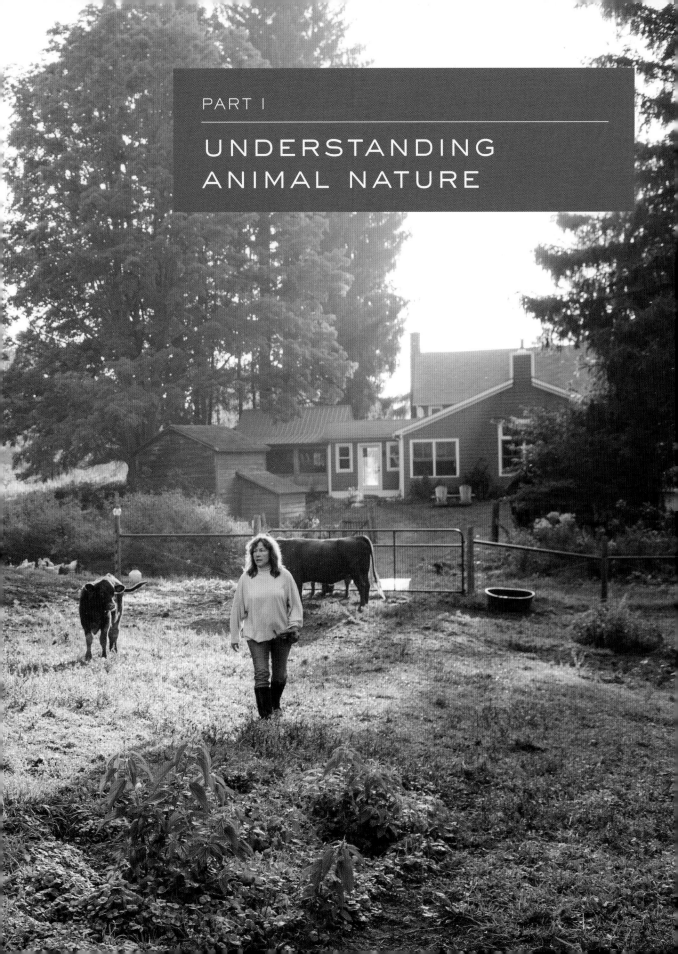

PART I

UNDERSTANDING
ANIMAL NATURE

1 HOW ANIMALS PERCEIVE THE WORLD

Animals are similar to us in some ways, and very different in others. People who become good stock handlers are especially observant for details of animal behavior that other people miss. They tune in to small changes, and a skilled handler can often induce calm animal movement with subtle motions of his/her own body.

Although all humans process information in a manner that is unique to their individual experience and genetic makeup, some adults think almost entirely in words, with very little visual imagery. Others of us can be characterized as visual thinkers. Like very young children, these thinkers process information in pictures based on experiences associated with their senses. Most humans combine varied degrees of verbal and visual ways of thinking.

Animals, however, think only in pictures, sounds, smells, touches, and tastes. This sensory-based thinking is highly specific, whereas the verbal thinking of humans is much more generalized. Verbal thinkers can easily overlook the sensory details that form an animal's world. This becomes a problem when people interact with animals based on their own perceptions, rather than attempting to view the world the way an animal experiences it.

PRINCIPLES

Vision and hearing are the primary senses animals use when vigilant or frightened; smell becomes dominant at breeding time.

Livestock will stay calmer if all yelling and whistling is stopped.

The first experience an animal has with a new person, place, or piece of equipment should be positive.

Animals never forget a frightening or painful experience.

How you touch an animal will affect how he reacts to you.

New things are scary when suddenly introduced, but they are attractive if the animal is allowed to approach them voluntarily.

Pigs have acute
hearing and
a powerful sense
of smell.

PIGS ARE NOT GRAZING ANIMALS, but many of their instinctive and learned behaviors are very similar to those of grazing animals. For example, pigs can be calmly moved using the same principles that are employed when moving the grazers. When we refer to grazing animals in this book, therefore, we include pigs unless otherwise noted.

Those who work with animals must start to see and hear what the animals are seeing and hearing. Animal handlers must become aware of how animals, ever vulnerable to attack in the wild, react to the human spaces they are entering, the sights and sounds surrounding them, the handlers' movements, and the emotional tones of the handlers' voices.

Making Sense of the World

Like all prey animals, grazing animals' sensory systems are adapted to be constantly vigilant. Unlike us, they are always on the lookout for predators, and their senses are sharp.

Grazing animals can hear higher-pitched sounds than people can, and they continually listen for signs of trouble. With eyes located on both sides of their heads, they can easily scan the horizon for danger as they graze. A sudden or novel sight or sound will cause a grazing animal to quickly raise his head. He will orient toward the stimulus, and his brain will make a decision: "Do I keep looking, run away, or put my head back down and continue grazing?" These unique sensory-system features greatly influence behavior during handling.

HEARING

All grazing animals, such as cattle, sheep, and goats, have very sensitive hearing and ears that move independently of each other. They often "watch" things with their ears. One ear may be oriented toward an approaching person or animal and the other to a passing vehicle. An animal's "ear radar" can indicate the things in the environment that the animal is noticing.

Ear position is also an important mood indicator. When animals pin their ears back, they are either frightened or aggressive. Alain Boissy of the French National Institute for Agricultural Research found that cattle also have a "relaxed" ear position and a "raised alert" position. Watch your animal's ears. They can tell you a lot about how he feels. Are the ears pinned, up and alert, or relaxed?

How an Animal Hears Your Voice

Animals are highly responsive to a person's tone of voice. They instantly recognize the voices of people who have been gentle and treated them with kindness, easily telling them apart from voices of handlers who have been abusive or have assisted in a painful procedure.

Livestock will stay calmer if all yelling and whistling is stopped. The only sound that people should make is a gentle *shhhh, shhh* with their mouths to move them.

In a herd or flock, at least one animal is always vigilant even as others graze.

Grazing animals "watch" things with their ears.

The best handlers are often completely silent when cattle are moving through the handling facility. When people learn to avoid loud, unnecessary noise they find that the animals remain calmer and are easier to move.

The tone of voice a handler uses around an animal is very important. Because animals have sensitive hearing, loud or high-frequency noises are frightening. Results from a study conducted by Jeffrey P. Rushen, a Canadian research scientist, show that screaming in an animal's ear creates a level of stress equivalent to the surprise jolt of an electric prod. In contrast, the calm voice of a known and trusted person can help soothe an animal that is scared.

Fine Tuning

Besides being sensitive to loudness and pitch, animals can distinguish between sounds directed at them by a person and noises emitted by equipment. Joe Stookey, an animal behaviorist at the University of Saskatchewan, has shown that the sounds of yelling and whistling increase the heart rate of cattle more than the sound of a gate slamming. Animals recognize the threatening intent of the yelling or whistling, which is unlike the neutral, and therefore safe, sound of the slamming gate. A more recent study done by Ruth Woiwode, of Colorado State University, showed that yelling, whistling,

CATTLE, SHEEP, GOATS, AND EQUINES can point both ears toward the same thing; or they can point their ears in two different directions and keep track of two different things. Here, the white goat and the Holstein cow both have one ear on the photographer. These animals have oriented their ears in order to hear any sounds from the camera more easily.

and running cattle during handling may reduce the animals' weight gain.

Animals make positive and negative associations with specific sounds or noises. For instance, they quickly learn the sounds of a truck bringing feed or a person calling them in to be milked or to switch pastures. A particular voice in the distance or the sound of a truck approaching or honking may be associated with the positive experience of being fed or scratched or brought into a safe and familiar place.

They can also distinguish between the sounds of two very similar vehicles or objects. They will approach a squeaky gate that goes to a feed room because they associate the sound with kindness or being fed and will avoid a squeaky gate that they associate with brutality or pain.

VISION

The most prominent of the senses governing the behavior of grazing animals, vision is the dominant one during handling. The location of the eyes on either side of the head allows a grazing animal to see nearly 360 degrees without moving his head. Such animals have only a small blind spot, in line with the backbone, directly behind their rear end. Their wide-angle vision enables them to scan the horizon continuously while grazing.

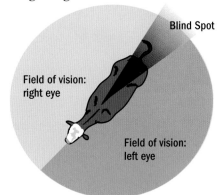

Blind Spot

Field of vision: right eye

Field of vision: left eye

THOUGHTS AND REFLECTIONS
Steps toward Good Stockmanship

Often I have stressed two essential things: the importance of numerical measurement of animal handling and the need for people to stop yelling. Many producers have adopted these principles. Doing this will prevent abuse and get livestock handling up to an acceptable level. The ultimate goal is to move beyond preventing abuse to creating more positive experiences for animals.

For numerical measurement, count any frightening or painful events that occur during handling. Examples include the number of percentage of animals that fall, hit a fence, are caught in the wrong position in a restraint device, or vocalize when you restrain them. Ruth Woiwode, a researcher at Colorado State, found that cattle that vocalize during restraint have a lower weight gain.

A good stockperson has a positive attitude and likes animals. Working to become calm and confident will turn you into an effective animal handler. Research by Paul Hemsworth and other scientists clearly shows that animals that fear people are less productive. Animals are very sensitive to your emotions. They can sense whether or not you are confident, too hesitant, fearful, or angry.

GRAZING ANIMALS have a horizontal band of extra-sensitive retina that makes it easier to scan their surroundings while grazing. Predators such as wolves and dogs have a central round area of maximum retina sensitivity.

more emotional right side of the brain. Initial careful presentation of a new thing, such as a saddle or trailer, to the right eye (left brain) may help reduce fear.

One study showed that tame cattle and horses that are relaxed around people may lose eye preferences. More research by Chelsey Shively and Mark Deesing showed that the direction of a horse's hair whorl affects handedness. It will determine which way a horse will bolt when startled by an umbrella suddenly opening.

Large ear tags can restrict vision behind the animal. This prevents the animal from being able to react to movements of people or animals behind it.

Grazing animals' vision is designed to detect motion, and with their constant vigilance they are aware of any movement that may signal the approach of danger. Slow, steady movement, on the other hand, in close quarters with the animals, is nonthreatening to them.

Left or Right Eye

Cattle and horses tend to look at novel, potentially threatening stimuli with their left eye. The left eye is connected to the

Bulls Can't Really See Red

Cattle, horses, sheep, goats, and other grazing animals can see color but lack the full spectrum of vision available to most humans because they have only two-color receptors. Human eyes have three-color receptors, which provide full-color trichromatic vision. Grazing animals lack the red receptor. They do not see red. They are most attuned to yellowish green and bluish purple hues.

Due to their two-color receptors, all grazing animals exhibit **dichromatism**,

Wool over Their Eyes

Sheep with long wool around their eyes are said to be "wool blind." These temporarily visually handicapped sheep may be more fearful or stressed during handling than shorn sheep because their visual field is reduced. Since they can't see around them, they sometimes don't know where a person is approaching from. When working with sheep, consider the length of the hair around their eyes and adjust your movements accordingly.

which means they are more sensitive to harsh contrasts between light and dark colors. Dichromatic sight enhances night vision and helps the grazing animal detect motion — but it can also lead to work slowdowns in an animal handling facility. Animals moving through a facility will often refuse to walk over a shadow or step onto a concrete floor from a dirt floor. The high contrast of the shadow or the flooring color change alarms them.

Depth Perception

To perceive depth, grazing animals must stop and lower their heads. If you watch cattle out in the pasture, when they walk up to a small creek or other natural obstacle they put their heads down to judge depth before crossing. The contrast between grass or dirt and water may look to the herd like a hole or ditch in the ground. The lead animal stops and lowers his head to see if the pathway is safe.

When cattle are moving through handling facilities, their instinct is to remain alert and keep their heads up. In this vigilant state, they may not want to risk putting their heads down to judge depth of shadows and contrasts in flooring. The lead animal may refuse to move forward or attempt to turn back if he is anxious about the situation.

Give Animals Time to Take in Visual Distractions

Try not to push the animals through a high contrast in flooring or past an unexpected change, such as a drain or a rubber mat. Instead, allow the leader time to put his head down and look things over. Often

TOP: Humans can see the color red, but cows see it as gray.
MIDDLE: In their familiar barn, animals will walk calmly over streaks of light they are accustomed to seeing.
BOTTOM: Animals may balk in this chute because the harsh shadows are new and novel. Sunny days may cause more problems than cloudy days.

it takes the animal only 30 seconds or so to determine that the footing is safe. After he has checked it out, he will walk over the shadow and the other animals will follow. If he cannot stop and look, he may get scared and turn back the herd in the opposite direction.

A good handler allows the leader time to look, especially when animals are being handled in a place they have never been before.

Ron Gill at Texas A&M University has a simple tip. The animal will be difficult to move when its nose is near the ground, it is looking at the ground, or its head is raised and vigilant. Wait for the head to return to the normal relaxed position before attempting to move the animal.

SMELL

Vision and hearing are the senses most often in use when an animal is frightened. Smell becomes dominant when a male is looking for a female in estrus; however, grazing animals do not rely on smell in the way dogs do.

A dog's social world revolves around smell. When he sniffs a tree that other dogs have urinated on, it is like getting the gossip at the local coffee shop. He knows who has been there, when they were there, their social rank, and whether the females were in estrus. A dog's world of smell, with its nuances and complexity, is difficult for people to imagine. Humans can and do develop exceptional ability to distinguish subtle differences among different smells. Wine stewards who can identify hundreds of similar wines by their smell probably experience the world more the way a dog does. To accomplish this, they must spend many years concentrating on their sense of smell.

Grazing animals can associate smell with something bad, although research in this area is limited. There is at least one case of a horse that was afraid of the smell of alcohol on a person's breath because someone who had been drinking beat him. A grazing animal's nose is at least as sensitive as a human's.

Some types of animals respond to a pheromone excreted from others of their kind who are terribly frightened. Observations indicate that stress has to be prolonged for the pheromone to be secreted. A single slip or poke with a prod is usually not frightening enough for it to occur.

I once saw an animal fall in a squeeze chute and get stuck for 15 minutes due to mechanical failure of the chute. The terrified animal urinated in the chute before he was finally released. The other cattle, even those who had not witnessed the fall, out-and-out refused to come near that chute until after the area had been washed down the following day. Experiments with cattle, pigs, and rodents show a common reactivity to the smell of the fear pheromone.

When thinking about animal handling, it is smart to consider that certain smells might cause an animal to behave with wariness or fear and to include that information as part of the puzzle of how to provide animals with the best possible experiences.

THOUGHTS AND REFLECTIONS
Try Shifting the Focus

I visited a place where they were working young cattle for vaccinations and ear tagging. No electric prods were being used and the people moving the cattle were silent. Yelling and screaming by people had ceased — but I still heard too much banging and clanging in the single-animal scale and squeeze chute.

A single small change — repositioning one person — caused most of the commotion to stop. This person, who had previously been stationed along the single-file chute, constantly waved his flag as he squirted grub medicine on the backs of the cattle. He was moved to the side of the single-animal scale that was located right behind the squeeze chute. He stopped waving his flag. I told him to apply the grubicide liquid in a line down the spine, moving from the animal's head to his tail. The movement of the applicator along the the back often motivated the animal to walk quietly into the squeeze chute.

TOUCH

How you touch an animal will affect how he reacts to you. Firm strokes are calming, but patting may be misinterpreted as hitting. Avoid light tickle touches because they may be scary.

The most effective calming touch should mimic a mother animal licking her young. Grazing animals are comforted by the feel of other animals around them, especially when they are closer than 3 feet (1 m) away.

Animals can make very specific touch associations. If a horse has had his mouth hurt with a jointed snaffle bit, he may become terrified when he feels a snaffle in his mouth. He may be perfectly relaxed wearing a one-piece bit instead because it provides a completely different touch picture. Violent or harassing touch will win a handler no friends or quiet cooperation in the animal world and should never be considered an appropriate way to influence animal behavior in a handling facility.

Sheep and cattle should be scratched under their chins and on their necks. Avoid touching or pressing on the forehead, which encourages butting.

In horses, a common problem is that the animal bucks when switching from a walk to a trot or a trot to a canter. Think about it — a saddle must have a different "feeling picture" at each different gait. It may have felt familiar and safe at a walk but totally new and scary at a canter. This problem is most likely to occur when training is done too rapidly and each new sensation is introduced too quickly.

TASTE

Animals can make very specific taste associations. Fred Provenza, a range scientist, learned that grazing animals prefer the feeds they ate with their mother when they were young and are more likely to eat less of an unfamiliar feed or forage.

Most cattle, sheep, pigs, and other livestock have a sweet tooth — just like humans do. Molasses, carrots, apples, and sweet horse-feed treats are effective rewards when you are teaching animals to cooperate for veterinary or other procedures. Take care, however, to offer small amounts only, or they will get sick. An animal that is 100 percent grass-fed is particularly vulnerable to bellyaches.

If you choose to raise goats, you will need well-built fences because these animals are escape artists.

Pigs need a diverse, well-balanced diet.

There are differences in animals' preferences for forages. Sheep and cattle prefer to eat grass. Goats are browsers and stand on their hind legs to eat tasty branches. All of these animals are ruminant grazers, and their efficient digestive system can break down tough cellulose plant fiber that humans, dogs, and many other **monogastric** (single-stomach) animals cannot eat. They have more than one stomach, and the huge first stomach, called the **rumen**, uses microbes to break down and ferment cellulose.

Equines — horses, zebras, donkeys, and mules — have a less-efficient system to break down cellulose. They use what is basically a giant appendix with a single opening for both the entrance and the exit of food. This poor design explains why horses have digestive upsets more often than ruminants do.

Pigs can break down some cellulose but not as efficiently as ruminants can. They are omnivores and can't live just on hay. They like to eat vegetables and, in the wild, even some small mammals.

Animal Emotions

The sensory-based thinking of animals is different from human thinking because it does not depend on words. Different sights, sounds, and other sensory stimuli can trigger reactions from different emotional systems in the brain.

Some people question whether animals truly have emotions, but research clearly shows that all mammals, including humans, have the same basic systems. Neuroscientist Jaak Panksepp has identified the primary emotional systems in the subcortex of animal brains, and those that motivate animal behavior are listed below. Each system operates as a separate brain circuit.

Fear (a negative emotion) is a primary emotional system that motivates animals to avoid predators. Fearful animals struggle during restraint, run away, bite, or kick during handling. It is important to differentiate fear-motivated behavior from aggression. Punishing fear-based behavior makes it worse.

Rage (a negative emotion) or anger enables an animal to fight off a predator.

Separation distress (a negative emotion). This emotion is distinct from fear. It occurs when an animal is isolated or when a mother and baby are separated. This is a separate brain system that is different from fear. Panksepp calls it panic, but I prefer the term **separation distress** to avoid confusing it with fear. When a single animal is separated from the herd and starts running around wildly, the separation-distress emotional system has been activated. Lambs bleat and calves moo when apart from their mother due to separation distress.

Animal Emotional Systems

Unfortunately, some scientists still question whether or not animals have emotions and consciousness. They think you need verbal language in order to be truly conscious. I completely disagree with this view.

One prominent scientist who questioned animal consciousness failed to review the evidence regarding how drugs such as antidepressants work in animals. In fact, Prozac works in dogs because they have the same neurotransmitter in their brains that people do. Another researcher, Gregory Berns at Emory University, trained dogs to lie still in the MRI scanner. When they smell their favorite person, the emotion centers light up, just as humans' do when they experience happiness.

When you stroke an animal, make your fingers feel like the mother's tongue. Do not use pats, which may be interpreted as hitting. Avoid light tickle touches as well because they are too alerting.

Seeking (a positive emotion, also called curiosity) motivates an animal to explore and investigate new things. Some animals are active explorers and will approach a novelty such as a lawn chair set in the middle of a field. Others are shyer, with less of an urge to explore. Genetic differences in the strength of the seeking trait are often obvious in dogs. A Labrador retriever bred for working in the field is an active, highly motivated seeker that wants to constantly chase a ball. A heavyset Labrador bred for service dog work is calm and not highly excited by balls.

Nurturing or caring (a positive emotion) motivates animals to care for their young. The hormone oxytocin is turned on during nurturing activities such as a mother nursing and licking her young and animals licking and grooming each other.

Lust (a positive emotion) motivates the sex drive.

Play (a very positive emotion) is enjoyed by all species of mammals, especially the young, such as calves and lambs frolicking and jumping out of the pasture.

Even though the emotional circuits in animals' brains are similar to people's, they may be wired differently in the body. A lack of awareness of these differences can cause misunderstandings between humans and animals. The most common example is hugging dogs.

Hugging around the shoulders is very positive for people, but it is not positive for dogs. When a dog puts its paw on another dog's shoulder, it is a sign of dominance. If you want to hug a dog, do it farther back on his body. Most dogs tolerate hugging, but they do not really like it. Their mouth will be clamped shut instead of partially opened and relaxed.

In farm animals, there are important differences in how they fight. Bulls and rams butt heads; horses rear up and strike with their front feet. Petting the forehead of bulls and rams will encourage butting. Avoid stroking or pushing on the forehead. Horses can be rubbed on the forehead because they are wired for rearing and striking instead of head butting.

Throughout this book, there will be more discussions of differences in behavior among species.

Sensory Thinking and Memory

How does a grazing animal understand the human world when his brain is filled with all these sensory details? Since animals lack language, visual images and the sensations of sound, smell, touch, and taste are stored in their memories. This is how they retain necessary information about past events in their lives.

Neuroscience research has shown that the animal brain puts sensory-based information into categories and stores it in separate files, similar to files on a computer. If an animal has only one scary or painful experience with a person, he is most likely to associate the bad experience with a single sensory-based feature of that person. Seeing a man with a beard may trigger a memory of being beaten by a man with a beard.

ANIMALS DO NOT FORGET

Sensory associations are an important part of an animal's learning process. If his first experience in a situation is bad, the animal may retain that visual fear memory for life. He may learn not to trust the place, the person, the equipment, or all of these things. On the other hand, certain experiences are associated with positive emotions. For example, he may associate the sound of a tractor with the positive experience of being fed.

Research by Joseph LeDoux, a neuroscientist at New York University and director of the LeDoux Lab Center for Neural Science, shows that fear memories in animals can never be erased. They can be suppressed, but if the animal is exposed to an event similar to the original, the memory may resurface. The fear memory file can never be deleted. When the animal learns to overcome his fear, the fear memory file is closed but remains in the computer's memory.

High-strung animals have a very sensitive trigger for retrieving fear memories. An Australian experiment using sheep illustrates this theory. Test sheep were restrained tightly in a squeeze chute and then turned upside down. When handlers tried to herd them into the same squeeze chute in the same place a year later, the sheep adamantly refused. They will remember the first frightening experience, even though a substantial amount of time has passed since the trauma.

MEMORIES ARE SPECIFIC

The images that animals process in their brains about their experience of the world produce amazingly specific memories, like exact pictures that are saved in a file for future use when needed. Animals use the stored pictures as a reference for how they should react to a new object, person, or place in the future. They form object- or person-specific fear memories and event- or place-specific fear memories.

Remembering People and Objects

The most common type of fear memories in animals are those that are attached to a certain object or person. I once observed cattle restrained in squeeze chutes for vaccinations and other veterinary procedures. During handling, a few animals were accidentally banged on the head with the stanchion that restrained them by the neck.

A black cowboy hat became inscribed in one horse's memory as a trigger of fear.

The next time the cattle were handled, they were taken to a squeeze chute in another facility that looked only slightly different. The animals that had been previously injured recognized the type of head stanchion and refused to enter it.

If the cattle had been taken to a squeeze chute that looked completely different, even in a strange place, it is likely that they would have entered willingly. The photographic memory the cattle created of the gate at their home ranch transferred to all similar head gates encountered from then on, regardless of the site. The original image of the head gate raised a lifelong fear in the animals.

I have also observed a steer that had on nose tongs for extra restraint in a squeeze chute. Nose tongs will cause pain for any animal. The next time the steer was worked, he entered the squeeze chute and head gate easily but tossed his head and reacted violently when he saw the nose tongs. (Because nose tongs are painful, if restraint of the head is required, I strongly recommend a halter instead.)

In another example, a man wearing a black cowboy hat mistreated a horse during a veterinary procedure. From that moment on, the horse was terrified of all black cowboy hats, while baseball caps and light-colored cowboy hats failed to incite a reaction. The horse's fear was specifically

Preventing Pain Associations

During dehorning, use of a local anesthetic will help prevent an association with the chute as the source of pain. When the anesthesia wears off, 30 minutes later, an animal is unable to identify the cause of pain and will not think of the head catcher as the source. This makes for far easier entry into the chute the next time around.

associated with the black cowboy hat his brain photographed during the abusive incident.

Two research studies verify the visual specificity of animal memories. An experiment showed that training a horse to tolerate the sudden opening of a blue-and-white umbrella did not transfer to the shaking out of an orange tarpaulin. A flapping tarp was novel and new and the umbrella was familiar and safe. Cattle that grew accustomed to being fed by people still became agitated the first time they were handled in the chute. The chute was a new, totally different "scary" thing.

In some cases of careless handling, abuse, or serious accident, an animal may be so traumatized he will absolutely refuse ever to reenter a handling facility or restraint, even when repeatedly shocked with an electric prod. The fear from that reawakened memory apparently outweighs the pain associated with an electric prod.

Ranchers who graze cattle in wolf country observe that cattle that have experienced wolf predation will develop an intense fear of dogs and may attack them. Working the cattle with herding dogs may become impossible.

Fear of wolves and dogs is learned. Ranchers report that cattle that have been attacked by wolves will not tolerate being handled with dogs. They may even intentionally attack old dogs. Oregon State University scientist Reinaldo Cooke found that cattle that had experienced wolf predation became more stressed when exposed to domestic dogs, wolf smells, and recorded wolf howls compared with cattle that had never known wolves. This shows the ability of an animal to remember previous frightening experiences. Wolves and dogs have many similarities, such as smell and appearance. A blue-and-white umbrella and an orange tarp, on the other hand, are totally different visually. Vision is the dominant sense when perceiving something that is scary. That's because in the wild, grazing animals depend on vision to detect threats before a predator gets so close that escape may be difficult. Sensory impressions can be one-dimensional (just one sense activated, such as vision), two-dimensional (vision plus hearing), or three-dimensional (vision, hearing, smell). As the impressions accumulate, the animal's fear grows.

This cow is curious about the cat and she cautiously approaches it.

Animal thinking and learning are very specific. Training an animal to tolerate one scary thing does not transfer to a new object that looks totally different.

Remembering Places

In some situations, the fear memory adheres to an object that an animal just happens to look at during a bad experience, instead of to the object or person that actually was responsible for the painful or frightening event. For example, dogs may fear the place where they were hit by a car instead of the car that hit them. This happens because the dog may not have been looking at the car when he was hit but instead, for instance, at the next-door neighbor's driveway. After the painful incident, it is the driveway the dog fears. Grazing animals often fear places where a frightening experience first occurred.

To the verbal thinker, this seems odd. A verbal thinker will expect the dog to be afraid of cars, not the neighbor's driveway. Most humans can reason this out because with our "higher-order" thinking, we have a greater understanding of cause and effect. Dogs and grazing animals, however, perceive danger in the way a four-year-old who has not yet developed cause-and-effect thinking might. Very likely, the animal will form a fear memory of what he was looking at when the accident occurred.

A horse was afraid of a barn where he had been whipped and had fallen down. Performing veterinary procedures in the barn was nearly impossible. When he was taken outside away from the barn, the procedure could be done easily. Another horse was calm when he was tied up with a single lead rope, but he got nervous and agitated when cross-tied, with a lead rope attached to each side of his halter. Two lead ropes were associated with an earlier accident, but one lead rope was safe.

Cattle Remember Specific Bad Places

Cattle retain specific memories of where a bad experience happened. Once I was moving some Angus and Hereford steers through a handling facility and up to the squeeze chute. Moving them in a wide alley through a crowd pen using flight zone principles (see page 66) was easy. When the calves got halfway up the single-file chute and saw the squeeze chute, however, they absolutely refused to move forward. They no longer responded to the point-of-balance principles, where the stockperson uses his or her position to move the animal. The animals stood frozen and started pooping. Frightened animals will often respond with loose manure and diarrhea. The flight zone had disappeared and I could put my hands on them, and they refused to move.

I did not know the handling history of these steers, but it was likely that they had a previous bad experience with having their heads banged by the head gate. When they finally entered the squeeze chute, the head gate was where they balked. They remembered the exact part of the squeeze chute that had hurt them in the past. After they exited the squeeze, their flight zone awareness returned and they responded normally by moving away. Once they were away from the squeeze chute, they knew that nothing painful or scary would happen.

The Physiology of Fear

Fear is an exceedingly strong stressor, and it can be worse than the pain from veterinary procedures. When animals are nervous and fearful a stress hormone, cortisol, is secreted, rising when they are handled roughly and diminishing when they are handled quietly (see graph below). Cortisol can lower immune function, and stressed animals are more likely to become sick.

In a number of studies, researchers have compared the cortisol levels of cattle, deer, and antelope in various stages of restraint. The animals with the lowest stress levels were tame, cooperative antelope that were given treats as they had their blood sampled. The levels of cortisol in wild cattle during routine handling in a squeeze chute were equivalent to the cortisol levels in gentle cattle while being branded. Cortisol levels of beef cattle that were handled calmly and quietly had only mild anxiety, two-thirds lower than those that were subjected to rough handling and use of electric prods. Similar findings have been reported in sheep.

A basic principle is that when animals are forcefully handled, stress levels are high. When an animal voluntarily cooperates, stress levels are low.

SIGNS OF FEAR AND DISTRESS

It pays to get to know the temperament of individual animals. This may not be possible if you work with a huge herd or flock, or if you house animals for only short periods of time. Still, however, you can understand much of what animals are feeling simply by observing their behavior.

CORTISOL LEVELS DURING RESTRAINT

This graph, compiled from various scientific studies, shows levels of cortisol measured in animal blood samples during various handling procedures. The animals with the lowest levels of cortisol in their blood, and thus with the lowest level of stress, behaved most calmly.

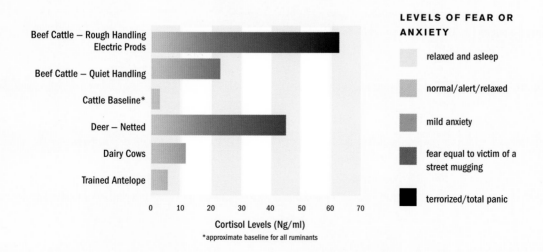

LEVELS OF FEAR OR ANXIETY

- relaxed and asleep
- normal/alert/relaxed
- mild anxiety
- fear equal to victim of a street mugging
- terrorized/total panic

Cortisol Levels (Ng/ml)
*approximate baseline for all ruminants

TAIL SWISHING;
REAR SOILED BY MANURE

EYE WHITE SHOWS

EARS MOVING IN OPPOSITE DIRECTION

HEAD UP

TAIL FLIPS UP

EARS PINNED BACK

GRAZING ANIMALS of all species exhibit similar signs when they are anxious, frightened, or panicky.

An animal will display one or all of these signs if he is experiencing fear or stress.

GESTURE	ANIMAL AFFECTED	SIGNAL
EARS MOVING IN OPPOSITE DIRECTION	All species of grazing animals	Each ear "watches" a different thing. At this stage, the animals are vigilant but not frightened.
HEAD UP	All species	The individual animal with the flightiest temperament will be in the back of a group. It will be the first one to put its head up.
TAIL SWISHING	Cattle, horses	Tail swishes when no flies are present. Speed increases as the animal becomes more frightened. Tail swishing is a warning before more violent behavior occurs. Not true for goats. A doe will wag her tail when she is in estrus (heat)
TAIL FLIPS UP	Bison, deer	First sign of fear. Note: goats naturally hold their tails up.
EARS PINNED BACK	All species	Frightened or angry
EYE WHITE SHOWS	All species, including dogs. Also called bug-eyed or wide-eyed	Sign of intense fear
DEFECATION	All species	Loose manure and soiling of animal's rear indicate extreme fear.

Teaching Animals to Trust

When animals trust a handler, fear and stress are nearly nonexistent. If calm handling practices are used during restraint in a squeeze chute or as animals are being moved into a new pasture, the animals will be more inclined to cooperate. When animals are roughly forced to do something, their stress will rise.

Previous experiences have a significant effect on behavior. At one dairy, the cows are accustomed to children playing in the milking parlors. At another, the cows became frightened because children were novel, new, and scary.

Teaching Animals to Cooperate

One advantage of a small flock or herd is that you can take the time to train your animals to cooperate voluntarily when you handle them. A tasty treat is a positive experience that outweighs the negative experience of getting an injection. On a large ranch, it is impossible to train every animal by feeding treats during procedures.

Animals deserve to live free from fear and pain, and treating them humanely should be a top priority for all handlers. The added incentive for **low-stress handling** is that it helps prevent injuries and keeps out-of-pocket health expenditures at a minimum. Practical experience shows that quietly handled animals are less likely to get sick.

MAKE FIRST EXPERIENCES STRESS-FREE

Put knowledge of animal thinking and memory function into practical use through actions that tell the animals their home environment is safe.

- Walk your animals through any new set of corrals and chutes and let them investigate on their own before any procedure is begun. This is especially important with young animals. Tie all the backstops and gates open to prevent banging and clanging that may frighten the animals.

- Create a positive association with new facilities such as corrals by putting out a favorite feed.

- Later, if animals must undergo a frightening or painful procedure in the corrals, take time afterward to reinforce positive associations with the area to prevent them from becoming afraid. (They may associate any pain or stress of the experience with some other new visual feature, such as an extra vehicle parked nearby, and attach their fear to it.)

- Take time to introduce new equipment in positive ways to elicit more cooperation from the animals.

- When transporting livestock, it is essential to drive carefully to prevent the animals from losing their footing and falling in a vehicle. Accelerate smoothly and avoid sudden braking to give your animals a safer, more comfortable ride. This pays off. In both cattle and sheep, two different studies showed that the first trip is more stressful than the seventh or ninth trip.

- When using a four-wheeler or other new motorized vehicle with cattle or other livestock, make sure that their first experience with it is positive. Allow your animals to approach and investigate. Even animals that are accustomed to a pickup truck will need to be carefully habituated to a four-wheeler or a motorcycle, because they will look and sound different.

Walk your animals through any new environment to give them time to get comfortable.

PREPARE FOR NOVEL EXPERIENCES

Animals find new experiences and novelties both attractive and frightening. Normal development of the brain requires new, unique, and varied sensory input. Novelty-seeking behavior is hardwired and guides the animal to explore his environment in order to discover new food sources and to find safe places to live. This is the seeking emotion that has been described by Jaak Panksepp (see page 22).

If you train livestock to tolerate different people, vehicles, and herding or driving methods, they will be less stressed when encountering new places and things. If you are showing your livestock or horses at a fair, for example, introduce them ahead of time to bikes, balloons, and flags. Never suddenly shove such fear-provoking items in your animal's face. The best approach is to allow your animal to investigate them voluntarily, long before the show. Decorate your pasture fences with balloons and flags and the animals will investigate them in their own time.

A cow perceives a person on a horse and a person walking as two different things, and she may react to them in two different ways. Most obvious is the difference in the flight zone, the point at which she turns to flee. A person on horseback may be able to approach as close as 6 feet (2 m) before she panics. When a person is on foot, the flight zone distance may triple.

In the brain of a cow, a man on a horse and a person walking are perceived as two different things, and she may react to them in two different ways.

Preparing Animals Not to Fear New Things

A study by Australian Geoffrey Hutson, a former senior research fellow at the University of Melbourne, indicates that offering tasty feed rewards to sheep after they leave the handling chute makes them more willing to enter the chute in the future. Reinaldo Cooke and his colleagues also found that acclimating animals to handling facilities and transport vehicles reduced stress and improved conception rates after artificial insemination. It is especially important to quietly acclimate young heifers by walking them through the chutes. On the first few trips through the chutes, they are not caught in the head gate. Limit to one training session each day.

Cattle used to people only on horseback can become dangerous to handle when first encountering people on foot. In a small confined area, they may attempt to run past the handler or race back and forth in an effort to get the person out of their flight zone. In a large pasture, they may scatter and run away.

If you are usually on a horse when you work with cattle, it is essential that they learn how to respond quietly to a person on foot who is moving them in and out of pens, in case they are moved to a new place where horses are not used during handling. Spend some time with them to carefully habituate them to a person on foot. Another method is to work on foot with a horse beside you.

NOVELTY TRIGGERS BOTH FEAR AND SEEKING

If animals are free to explore on their own, novelties are attractive and the animals' curiosity is heightened. Suddenly faced with a new object, on the other hand, especially in a confined area, they will most likely react with fear. If a paper bag is left inside a pen, for instance, the animals will approach and investigate it at their own pace. But if the wind suddenly blows a bag in, they will be startled and jump away. High-strung animals are the most sensitive, being the first to investigate and the first to become frightened and flee. They experience more extreme emotions on both ends of the attraction and fear spectrum.

Studies show that behavioral reactions to novelty decrease with repeated exposure. If you train your animals to accept the unfamiliar they will be less likely to become severely stressed when taken to another ranch or facility, or the processing plant.

TO HAVE RESILIENT ANIMALS, VARY THE ROUTINE

Animals raised on a single ranch or farm for their entire life are not exposed to enough variation. Typically, the same

I have observed grassfed cattle that have been allowed to go "feral" because nobody has worked with them. It is the responsibility of a good stockperson to never let their livestock turn into "wild animals" because nobody handled them for two or three years. This animal is afraid of the novel roller machine. Note his ears: one is pointed toward the roller machine and the other toward the photographer.

Some animals are active explorers and will approach an unfamiliar object in their own time.

person feeds them from the same truck at the same time every day; movement from pasture to pasture becomes routine. These animals become accustomed to the routine and are considered calm and easy to work with. The owners of animals like this are often surprised when they take the animals to the auction yard or packing plant and they balk at everything in their path and attempt to jump fences.

Many stockpersons believe that it is best for animals to have a routine. The time of day for milking or feeding should indeed be the same every day; however, livestock must learn to be comfortable with a variation in routine as well, regarding things they will see or hear. Pigs, for example, will be less likely to become agitated when hearing a new sound if a radio is kept playing at a reasonable volume, tuned to a station with a variety of voices and music. In the same way, teaching animals to tolerate some variation in vehicles and people who handle them will help make them less fearful of new experiences.

Introduce animals to strange and unusual objects, different vehicles, and new people, and allow them to investigate them on their own. They need to look at and sniff new things before accepting them. Not all animals will react the same way to things. Something scary and new for one animal will be neutral or possibly attractive to another animal, so it is important to accommodate these differences. Each will respond to new stimuli in a manner dictated by his previous experiences or inexperience with similar settings, and by his individual temperament.

When you sit quietly, cattle will come up to you to get acquainted.

Fearful or Curious?

Scientific researchers have discovered a "switch" in the brain that can trigger either the seek or the fear mode. Called the **nucleus accumbens**, this switch explains the behavior of an animal as it orients toward a novel sight or sound. During the orienting phase, when the animal raises its head and focuses on a stimulus, the brain can either remain in seek mode and keep watching or switch to fear mode and initiate running away. When the switch is in neutral, the animal will resume grazing.

The brain switch mechanism can be biased toward either fear or seek, depending on both genetics and previous experiences. Animals that were abused at a young age will have a stressed nucleus accumbens that can switch more easily into the fear mode. There is also evidence that abuse that occurs during pregnancy can up-regulate the nervous system of the fetus and make the offspring more fearful.

The position of the hair whorl on this animal's face indicates an even temperament.

2 GENETICS AND LEARNED BEHAVIOR

Behaviors can differ greatly among species. Cows and goats, for example, isolate themselves when they have their calves or kids. Ewes, on the other hand, will give birth to their lambs close to the flock. Genetic factors determine instinctual behavioral patterns that are unique to each species.

Genetics can also affect the strength of the basic brain emotional systems (described in chapter 1), which are the same in every species. Animals can be bred to be high or low on the seeking trait, or high or low in separation distress. Individuals with high separation distress will become more agitated when they are isolated from the herd, while others remain calm when on their own. Research now shows that many traits have variable levels of expression.

Another important factor in shaping the behavior of livestock is life experience. The early experience of any animal — his mother's care, his health, and his treatment by humans and other animals — determines how he will respond to people, animals, places, and objects later in life. Many people have a pet or a favorite animal, and they can perceive distinct individual behaviors that make up its personality. Both genetics and experience determine these individual character traits, such as sociability and the animal's degree of reactivity or fearfulness when he is suddenly confronted with a sight or sound he has never before experienced.

PRINCIPLES

Never overselect for a single appearance, production, or behavior trait.

Overselecting for a single trait may cause problems that are difficult to predict.

Select for optimum production, not maximum.

Animals selected for a calm temperament are more productive.

Bulls hand-reared as orphan calves may be the most dangerous because they have not learned that they are cattle.

Never Overselect for a Single Trait

Ranchers and other animal producers can purposefully select animals to breed and raise. Selection can be based on growth rate, body size, temperament, muscle enhancement, disease resistance, and other factors influenced by genetic code. Even conditions such as susceptibility to lice infestation are influenced by genetics: some cattle will get a worse infestation than others. It's important to note, however, that none of these traits are mutually exclusive. When choosing to breed animals for a single behavior or body feature, other traits will likely emerge or diminish.

WEIGHT GAIN VS. DISEASE RESISTANCE

A big mistake made by many producers is to overselect for a single trait, such as weight gain, carcass quality, temperament, or appearance. This will cause big problems, since there are always trade-offs. Animals that gain the most weight, for example, might lose disease resistance. Researchers studying Scottish Blackface sheep in England found another genetic trade-off between different traits. Sheep with higher reproductive function had lower immune function, and sheep with higher immune function had fewer lambs.

In 2013 and 2014, huge numbers of high-producing commercial piglets and poultry were lost due to porcine epidemic diarrhea and avian influenza. To reduce the possibility of another big epidemic, farmers should examine the influence of genetic factors on disease resistance. If you are raising organic animals, I strongly recommend that you select hardy, slow-growing, disease-resistant genetics.

Animals such as these Scottish Blackface sheep need balanced genetics in order to thrive.

The Music Mixing Board

An easy way to understand how genetics affects animals' emotional behavior is to picture a sound mixing board. The mixing board has sliding volume switches for many different channels. Imagine that each sliding switch controls the volume of a separate emotional trait: (1) fear, (2) rage, (3) seeking, (4) separation distress, (5) sex drive, (6) nurturing, (7) play. Each emotional trait can be adjusted for intensity (high, medium, or low) through breeding, and some breeders assign numbers to identify the strength of the trait.

STRUCTURAL DEFICIENCIES OF THE FOOT AND LEG

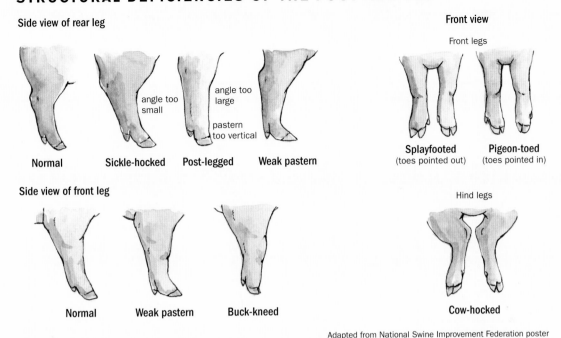

Side view of rear leg

Normal • Sickle-hocked (angle too small) • Post-legged (angle too large, pastern too vertical) • Weak pastern

Front view

Front legs

Splayfooted (toes pointed out) • Pigeon-toed (toes pointed in)

Side view of front leg

Normal • Weak pastern • Buck-kneed

Hind legs

Cow-hocked

Adapted from National Swine Improvement Federation poster

CARCASS QUALITY VS. STRUCTURAL SOUNDNESS

Some breeders choose sires and dams solely by looking at the numbers for traits such as growth or carcass quality. Visual appraisal is still essential, though, to stay out of trouble. It is especially important to select breeding stock with sound feet and legs.

Overselection for specific carcass traits may be related to an increase in leg conformation problems in cattle and pigs. In the 1980s, with the introduction of rapid-growth pig genetics, problems with poor leg structure and lameness started to emerge. More recently, beef cattle overselected for good carcass traits have had similar problems with poor leg conformation and lameness.

A leg conformation chart can be used to select animals with structurally sound feet and legs. The chart above was developed for pigs, but cattle, sheep, and other animals can have the same conformation problems, which can contribute to lameness. Common defects include post legs (too straight), twisted toes (corkscrew claw), crooked legs, and collapsed pasterns. Another defect is twisted claw or corkscrew foot. As the animal ages, the toes become more and more crossed.

Select for Optimal Production

Traits are linked in unexpected ways. A good rule of thumb is to select for *optimum* production, not maximum. It is a mistake to select animals strictly by their production numbers. There will always be a need for visual appraisal of animals to help producers breed animals that are structurally and emotionally sound.

When breeding animals, producers must not allow "bad to become normal." Bulldogs are a good example. Today, people who are first introduced to the breed think it is normal for bulldogs to have breathing, joint, and birthing problems. Photos from the 1930s to the 1950s of sports teams with a bulldog as the mascot show a functional animal with less extreme appearance traits than those of bulldogs today. The same principles also apply to farm animals.

MILKING BREEDS VS. MEAT BREEDS

Both cattle and goats have been bred for either milking or meat. Specific milking breeds include Holstein, Jersey, and Guernsey cattle and Alpine, LaMancha, Nubian, and Toggenburg goats. These animals usually have a slenderer body than breeds selected for meat, such as Angus, Simmental, and Hereford cattle and Boer, Kiko, and Spanish goats, which are heavier and more muscular. There are also dual-purpose breeds that are used for both milk and meat.

ADVANTAGES OF CROSSBRED ANIMALS

In the world of breeding, as throughout the natural world, diversity is a good thing. When two different breeds are crossed, the offspring is called a **filial 1** (F1). An F1 offspring has a dam and a sire of two different breeds. Breeding divergent (highly different) genetic lines of animals often results in **hybrid vigor** and a much lower incidence of inherited defects. Hybrid vigor is a scientific term that means the offspring may be stronger than the parent and have many desirable traits.

These traits may be lost, however, if there is continued breeding of animals that are too closely related. An example of closely related breeding that can get a breeder in trouble due to inherited defects is breeding fathers to daughters or brothers to sisters. Sometimes closely related animals are mated to strengthen desired traits, but it has to be done extremely carefully. People interested in animal breeding should consult genetics books, experienced breeders, or their Cooperative Extension Service.

The Importance of Preserving Rare Breeds

The Livestock Conservancy lists rare breeds based on their risk of being lost. In cattle, the breeds at greatest risk are mostly dairy breeds, including the Milking Shorthorn. I can really relate to Milking Shorthorns because I learned to hand-milk them when I was a teenager. Other endangered dairy breeds include Ayrshires, Galloways, and Guernseys.

Among sheep, there are valuable rare breeds that are hardy and parasite resistant. (Internal parasites are a big problem in sheep.) Heritage wool breeds include the American Karakul, the Navajo Churro, the Lincoln, and the Romeldale. Both the Gulf Coast Native sheep and the St. Croix are hardy, parasite-resistant animals that can tolerate high heat and humidity. Shropshires are a dual-purpose sheep with a gentle disposition.

Hereford pigs, with the same color markings as Hereford cattle, are docile and easy to raise.

MILKING SHORTHORN

LARGE BLACK

HIGHLAND CATTLE

ST. CROIX SHEEP

HEREFORD PIG

SHROPSHIRE SHEEP

According to the Livestock Conservancy (livestockconservancy.org), based in North Carolina, heritage breeds have better mothering ability and are hardier and healthier than industrial breeds.

BODY SHAPE AND TEMPERAMENT

A gene makeup that encodes an animal's body to grow large and stocky, regardless of breed, also often produces an animal that is calm and less likely to become fearful. On the other hand, animals with a fine-boned skeleton, little fat, and a long slender body often have a highly reactive temperament and are flightier — easily startled and frightened. This relationship between body structure and temperament has been observed in many animal species, such as cattle, pigs, and chickens.

TRAITS ASSOCIATED WITH NEUROLOGICAL PROBLEMS

A relationship between coat color and temperament can be found in many animal species. Dairymen comment that mostly white Holsteins are flighty and more dangerous to handle. An orange marmalade cat is often very affectionate but skittish. Animals with too much of the albino traits, such as light-colored paint ponies, are beautiful but have a reputation for real flightiness. There are often neurological problems with animals with white coats and blue eyes. In both horses and dogs, breeding two blue-eyed animals together may result in deafness, epilepsy, or abnormal behavior.

Animals with white coats and pink skin may be partially albino and are more likely to have problems. An Arab horse or Brahman cow with a white or light gray coat and dark-colored skin, on the other hand, will usually have normal nervous system function.

Breeding for Behavior

Within a breed, the behavioral traits of animals can vary. Some animals may be flightier than others. On average, Hereford cattle are calmer than Salers, but there may be a few Herefords who may be similar to some of the calmer Salers. Behavior traits are heritable and producers can select for them. In wild animals there are differences in temperament; some individuals will be more fearful than others.

GENETICS AND CATTLE BEHAVIOR

Domestic cattle breeds are traditionally split into two groups: Continental European and British breeds (*Bos taurus*) and breeds from India (*Bos indicus*). In the United States, the most common of the *Bos indicus* breeds are Brahman cattle. They are useful in hot regions because of their ability to withstand heat and flies, and they exhibit hardiness against diseases to which European breeds are susceptible.

Cattle breeds developed for high-quality meat when fattened on grass pastures and those selected for fattening on grain are very different animals. The British breeds (such as Hereford, Devon, Angus, and Shorthorn) were originally developed for grass-fattening, and they are usually calm and less flighty. The Continental breeds (such as Limousin, Simmental, Salers, and Charolais) have greater amounts of lean muscle, but some genetic lines are known to be nervous and flighty.

Producers who grass-fatten cattle will have better-quality meat and calmer animals if they choose either older genetic lines of British breeds or other breeds with similar characteristics. These animals will be smaller and have more marbling in their meat than those with lean genetics that were selected for fattening on grain. Although they are all beef cattle, their genetic makeup manifests in different body types and temperament.

As a comparison, consider one of the most popular of man's best friends, the Labrador retriever. Many people are familiar with lean Labs that want to be outside every day galloping across fields and leaping high into the air for tennis balls. Equally familiar are the bulky Labradors that are quite content to lie in the shade or on the porch all day. These calm, heavyset dogs make wonderful, patient service dogs. They are both Labrador retrievers, but because they have different genetic lines, they look and behave differently.

Temperament Issues

Cattle breeders sometimes observe aggressive temperament problems in crossbreeds between certain genetic lines of European and Brahman breeds. The resulting offspring sometimes lose a desirable trait of one of the parent breeds. Certain genetic lines of purebred Brahmans are very affectionate when they are handled quietly. They will often seek stroking from people. When Brahmans are bred to a *Bos taurus* breed, the affectionate trait may be lost.

Handlers and producers should be aware that there may be problems with either purebred Brahmans or Brahman crosses becoming extremely agitated and fearful when they are suddenly

Brahman cattle line up and look at the photographer. The animals with more excitable, flighty genetics stay back behind the others.

confronted with the novel surroundings of an auction yard or slaughter plant. These same animals may have been very calm in the familiar home environment.

A Brahman may become either immobile or aggressive when frightened by a new environment. When he becomes fearful, he may lie down in a chute or truck and refuse to move. When this happens, it is best to leave the animal alone; when he calms down, he will get back up.

Some flighty genetic lines of European breeds and their crosses may have a different response. They may instantly thrash or flee, injuring themselves in the process. In other words, one genetic type responds to fear by freezing, and another type has an instantaneous flight reaction.

In response to these problems, some breed associations have added temperament testing to their selection criteria. Even with the vast improvements in handling methods, there is a need to select for temperament. Animals with a highly reactive temperament that have been handled quietly at home often become extremely fearful when brought to a scary new place, unless they have been carefully trained to tolerate many new things and people at home.

In some breeds of cattle, temperament selection has been continuous since the early 2000s. Many lines of purebred cattle are calmer compared to 20 years ago. There is a concern, however, that overselection for super-docile animals may reduce desirable traits such as foraging ability or defense against predators. It is best to get rid of the really crazy animals instead of selecting for the calmest and least-reactive animals.

GENETICS AND PIG BEHAVIOR

There are significant differences in the behaviors of the different breeds of pigs. Some hybrid pigs that have been selected for rapid lean-muscle growth are often more excitable than older genetic lines of purebreds such as the Yorkshire. The more excitable types of pigs are more likely to pile up and bunch tightly when they become frightened. C. L. Yoder and colleagues at North Carolina State University found that pigs that remain calm during handling gain weight faster than those that become agitated.

Another difficult behavioral trait of some genetic lines of lean pigs is increased aggression and fighting when strange pigs are mixed together. This can cause serious problems in group housing. It is best to choose less-aggressive pig genetics.

SHEEP BREED TEMPERAMENTS			
STRONGLY GREGARIOUS (Tight flocking behavior)	**MODERATELY GREGARIOUS**	**WEAKLY GREGARIOUS** (Loose flocking behavior)	**SOLITARY**
Merino	Columbia	Barbados	Bighorn
Navajo	Finnish Landrace	Cheviot	Corriedale
Panama	Montadale	Dorset	Lincoln
Rambouillet	Polypay	Hampshire	Suffolk
	Targhee	Oxford	Wild sheep
		Shropshire	
		St. Croix	
		Suffolk	

Adapted from *Sheep Production Handbook* (Centennial, CO: American Sheep Industry Association, 2002).

GENETICS AND SHEEP BEHAVIOR

Sheep breeds differ significantly in their instinctual flocking behavior. Some breeds will flock together at the smallest sign of a threat. The fine wool breeds, for example, are often more gregarious and will flock more tightly. These breeds are easy to handle and will move together in any direction you wish with very little effort. Other breeds are more independent and the instinct to form a tight flock is weaker. Australian researcher Victor R. Squires classified the flocking behavior of different sheep breeds into four categories of gregariousness (see table, above).

GENETICS AND GOAT BEHAVIOR

Compared to sheep, goats are more curious; they like to climb on fences and escape. While sheep will stay flocked together, goats are independent. Goats have great personalities and will often quickly come up to the fence for attention.

A strikingly unique behavior in one goat breed is that of the myotonic, or "fainting," goat. A sudden noise will cause these goats to stiffen up and fall over. They remain conscious and will get back up when their muscles relax. (In people, a similar genetic condition causes muscle stiffness.)

Formally recognized as a breed by The Livestock Conservancy, myotonic goats are popular because they have so many wonderful characteristics. They are extremely muscular and are **easy keepers**, meaning they require less feed to maintain body condition than comparable animals do. Desirable behavior traits of myotonic goats are docility, good maternal behavior, nonaggression, and parasite resistance; these goats are also less likely than those of other breeds to jump up and escape from the pens. My research convinces me that they are a great choice for somebody who wants a docile goat for a child to raise.

SHEEP	GOATS
Flock together and follow easily through single-file chutes	Are more independent than sheep and move through handling facilities less easily. Leading goats may be more difficult.
Herd easily with dogs; flock together when frightened	May work poorly with dogs because they tend to run when frightened
Usually do not climb fences	More likely to climb fences. Goats are more agile than sheep and will often stand up with their front feet on a fence to look around.
Normal tail posture is down.	Normal tail posture is up. Ewes will wag their tails for the buck when in heat (estrus).
When lambing, sheep do not isolate themselves.	When kidding, goats will isolate themselves (lying out).
Prefer to eat grass; only 10% of diet is browse	Prefer to eat browse, which makes up 60% of their diet
Rams fight by lowering their heads and backing up before charging.	Bucks fight by rearing and quickly moving toward the opponent and butting him.
Rams have no odor during mating season.	Bucks have a strong odor and will scent-mark by urinating on their beards. This is normal.

Individual Differences

Parents of an only child often believe that the child's behavior, personality, performance in school, and other traits are an exclusive result of upbringing. It's when they have a second child that they start to understand individual differences. Even when they raise the second child exactly the same way as the first, unique behaviors and personality types emerge.

Like humans, animals can be reared under the same conditions but display considerable variations in behavior. In a litter of puppies, some will be shy and nervous while others are outgoing and bold. Psychologists and animal behaviorists understand that within any population, individual temperaments will range from one extreme to the other, with the majority somewhere in the middle. Anyone with lots of experience raising animals will find obvious behaviors that range from the extremely fearful and shy to the extremely fearless and outgoing, even among animals that to the untrained observer appear to be behaving exactly the same.

The basic dimensions of temperament are well understood. The temperament trait we call fearfulness greatly affects an animal's reaction to a sudden novel experience. The animal with the calmer genetics will walk around sedately at an auction, and the animal with the high-fear genetics may crash the ring fence.

Be aware that excessive fear and anxiety may reduce animal welfare and performance. For instance, fear reactions in livestock, such as agitation and violent escape, are often inappropriate in intensive systems. Highly fearful individuals can influence the behavior of the entire group. When grazing animals become frightened, a wave of escape behavior can pass through the herd, resulting in big group pileups (not to mention the possibility of broken fences and stampeding runaways). Animals can perceive the stress state of their herd mates, and the fear contagion will spread.

Early Experience and Temperament

The moment a newborn animal opens his eyes and takes his first breath, his behavior begins to be shaped both by genetic factors and by environmental circumstances. Genetic factors predispose him to respond to the environment in particular ways. An animal born with a flighty, fearful temperament will react differently than an

animal born with a more stoic nature. With that in mind, producers can thoughtfully control the environment and early experiences of their animals and thereby shape the subsequent behavioral reactions to people.

Here are some guidelines for getting babies off to a good start.

Handle young animals carefully and gently. As discussed in chapter 1, if you use rough handling methods when your animals are young, strong fear memories can form negative associations toward people in general. Making the young ones' first few direct experiences with people calm and quiet will keep them feeling safe and cared for in their earliest and most vulnerable days and make them easier to handle when they grow up.

Spend time quietly walking among the herds from birth to weaning so that the new calves, kids, and lambs learn that humans aren't a threat. The young animal learns fear or anxiety from his mother. If the mothers are calm while a person is walking among them, their offspring will follow their example and remain calm. Research with sheep has shown that the presence of a familiar person can reduce stress caused by isolation. A familiar, safe person may also reduce the stress of being separated from the mother at weaning.

Let animals have a good first experience with new objects. Introduce new things, such as minivans, four-wheelers, tractors, and unfamiliar people, to young animals in a nonthreatening manner. If the first

time they see a four-wheeler it chases them, the animals may fear it for life. Tip: A good way to introduce a four-wheeler is to use it to bring food.

TAGGING AND CASTRATION

Early experiences with humans can have profound effects on the way animals perceive people thereafter. It is unknown, however, if handling calves or other animals for ear tagging or castration shortly after birth has strong negative effects. Animals this young may not have the capacity to form associations and develop fear memories. Even calves that are a few weeks old may not remember bad experiences, whereas at several months old they have an excellent memory for both good and bad.

Practical experience suggests that performing these procedures as soon as possible after birth reduces the chance that the calf will remember and associate people with the pain. Even so, the animals must receive painkillers and anesthetics during these procedures. Even though a very young calf may not consciously remember a painful experience, there may be a lasting effect on his nervous system reactivity, making him more fearful of future stressors.

REDUCE WEANING STRESS

When possible, use low-stress weaning methods. For all species, the lowest-stress method of weaning is to allow the young to gradually wean themselves. If the young have to be weaned abruptly, it is best to leave the calves, piglets, kids, or lambs in a familiar pen or pasture. This avoids the double stress of separation from the mothers and introduction to an unfamiliar place at the same time.

Walk quietly among the youngsters and give them a feed they like to reduce stress during the first few days after weaning. It is especially important at this time not to move quickly or yell.

Both practical experience and scientific studies show that vaccinating and weaning calves 45 days before they leave the ranch greatly reduces stress and sickness when they arrive at their destination. Some veterinarians recommend 60 days. Vaccinating calves on the same day they leave the ranch is not recommended. Vaccines need time to trigger an immune response before the calves are exposed to strange animals at their destination.

The worst way to wean is to suddenly separate the unvaccinated calves from the cows and put them in a truck. This causes the distressed calves to pace and constantly bawl after they are unloaded at their destination.

To reduce stress, use a low-stress weaning method such as fence line weaning or nose flap weaning, a method used extensively in South America. Both of these methods, described below, allow a calf to stay close to his mother and will drastically reduce bawling and pacing. Cows and calves are more concerned about being close together than they are about nursing.

Fence Line Weaning

In fence line weaning, cows and calves are separated by a sturdy fence but can still touch noses. When calves and cows can still touch each other, even if through a fence, high-pitched distress moos are reduced. By the time the calves are weaned, they no longer need milk for nutrition and can get all they need from grass. Some pace and walk the fence, but there is less of the bawling and stress that result from sudden and complete separation from Mom. After a few days, the animals gradually move away from each other. Fence line weaning will *not* work for bison.

Fence line weaning allows cows and calves to remain close together. The calf is more interested in staying close to his mother than in nursing from her.

Nose Flap Weaning

When weaning calves, you can clip a plastic flap in their noses; the flap serves as an anti-sucking device to prevent them from nursing. Once the flaps are in place, the calves can return to their mothers for between 4 and 14 days. After this period, you remove the flaps and the cows and calves are permanently separated from each other. There is some pacing and mooing, but this weaning method, next to natural weaning, causes the least upset.

Some tips for success with this method:

- When nose flaps are used, calves must have either high-quality pasture or supplements to ensure adequate intake of nutrients.

- Nose flaps are not recommended in situations where pasture is poor unless the calves have access to supplements.

- Use high-quality flaps that have no rough edges to hurt the nostrils. Avoid designs with spikes because they may make the mother reluctant to nurse her next calf.

One downside to nose flaps is that producers have to put the animals in the squeeze chute or on the calf table twice: once to put the clip on, and again to take it off. If you're using a vaccination program that requires you to vaccinate your calves twice, it's possible to install and remove the flaps when the vaccinations are given.

Temperament-Testing Your Animals

To get a handle on what sort of individuals you have in your herd, you can conduct temperament testing. This process will help you identify highly fearful animals that are likely to be scared of new things.

You must evaluate temperament more than once for an accurate assessment. In a study my colleagues and I conducted on the temperament of nearly 100 bulls, the animals were handled four times at 30-day intervals. When handled, 9 percent of the bulls became highly agitated in the squeeze chute and 50 percent remained consistently calm. The remaining 41 percent were sometimes agitated and other times calm.

It was important to have several evaluations to accurately identify the 9 percent of bulls with the worst (most excitable) temperament. If you plan to cull such animals from the herd, multiple evaluations will provide a more realistic assessment of individuals than will a single evaluation. This will help you avoid culling an animal that became agitated only because an animal next to him became excited.

Measuring speed out of the squeeze chute is a good way to assess temperament in grazing animals. Studies conducted at Texas A&M University using a laser device to record speed show that exit speed is highly correlated with higher cortisol levels and lower weight gain. The faster the exit speed, the higher the cortisol levels.

EVALUATING CATTLE

Calf exit speed scores measured at weaning were the best indicators of the temperament of calves as adults. Testing calves at weaning is also more accurate for measuring genetic differences because calf behavior is less influenced by experience than adult behavior is.

It's important to note that adult animals that have become trained to low-stress handling methods may not show their true genetic reactivity during a temperament test in a familiar handling facility. An animal with innate high fear may go berserk when taken to a new place because he gets frightened more easily in a novel environment.

Another method for temperament testing involves exit-scoring animals based on traditional horse gaits (walk, 1 point; trot, 2 points; and canter, 3 points) to measure agitation coming out of a squeeze chute. This is more accurate for revealing differences among breeds of cattle or differences among genetic lines within a breed than scoring behaviors of cattle held tightly in a hydraulic squeeze chute. To find out which genetic line is the calmest, simply score the different types by adding up the points that represent how fast each animal belonging to that breed comes out.

Another behavioral trait influenced by genetics is sociability. (This may be the same as Jaak Panksepp's nurture trait; see page 22.) Some breeds of cattle are much more responsive to stroking than others. Purebred Brahmans that have been treated gently are often more responsive and will seek out stroking more than a Hereford will, for example.

Rate Livestock Temperament

A simple temperament-scoring system can be used to understand your herd better and to make decisions about breeding and selection. Rate each animal during restraint in a manual squeeze chute, head gate, or single-animal scale. Record your results. Score your livestock periodically to examine if your herd is getting calmer or wilder.

Note: This method may be less accurate if an animal is very tightly restrained in a hydraulic squeeze chute, which greatly restricts its movement.

IF THE ANIMAL IS:	SCORE HIM:
CALM	
Standing still	1 point
RESTLESS	
Intermittently shaking the chute	2 points
STRUGGLING	
Continuously shaking the chute	3 points
ACTING BERSERK	
Making wild attempts to escape	4 points

Using the Results

When you make decisions about culling and breeding, don't select only for those cows that repeatedly score 1 point. If you do, their offspring will likely be calm, but they may refuse to mother their babies. Hotheaded cows that run off, break gates, and repeatedly display disruptive behavior should be culled; three strikes and they're out. Bulls that even think about attacking someone should be culled immediately.

Striking a Balance

While behavioral agitation in the squeeze chute measures fearfulness (mentioned earlier), a temperament test used on the Lasater Ranch in Colorado assessed both sociability and fear. The test entailed people sitting on bales of hay and holding out sticks with a yummy treat on the end for weaned calves to sample. The calves were scored for sociability by determining their willingness to approach and eat the food. This producer culled any animals that didn't approach the stick, and the result was a ranch full of calm and friendly cows and very gentle bulls that you could scratch without fear of attack. To maintain hardy genetics and good mothering, any cow who failed to give birth to and raise a calf until it was weaned was also culled.

Ranchers have found that they can select cows that are gentle toward people yet still able to defend their calves against coyotes. Their three selection criteria are very simple:

1. The cow must wean a calf every year.

2. She must breed back quickly.

3. She must not be aggressive toward people (if she is aggressive, she is culled).

These criteria may decouple the vigilance and fear traits. The selection methods on the Lasater Ranch, and the three criteria above, result in cattle with higher hair whorls (see next section),

heavier-boned frames, and tolerance for people handling their calves. These ranchers have bred a unique animal.

Researchers in Mexico studying Zebu cattle have found that the ability to defend a calf is a separate behavior trait, not associated with a standard temperament test measuring degree of fearfulness.

Hair Whorls and Temperament

Since the late 1990s, people have been selecting cattle to have a calmer temperament. This has resulted in fewer cattle with a hair whorl high up on the middle of the forehead (see page 51). Many cattle today have a hair whorl either slightly above the eyes or way below the eyes. When I looked at old pictures of beef cattle from the 1970s and 1980s, there were more animals with high whorls in the middle of the forehead than are found today.

A study by Cornelia Flörcke on a Colorado ranch found that a higher hair whorl is associated with vigilance. Among her findings:

- When a strange vehicle threatened a mother cow and a newborn calf, the cows with a high hair whorl looked up when the vehicle was farther away than the cows with a lower whorl did.

- Some mothers moo (vocalize) to call their calf while others do not.

- Some Angus cows that have no whorl but have plain straight hair on their forehead were more likely to lose a calf to predators.

Cattle Hair Whorls

I first met Mark Deesing in the early 1990s when he came to my office at Colorado State University to discuss his theories about the relationship between the position of the round spiral hair whorl on a horse's forehead and his temperament. Professors at several other universities would not listen to him, but I did.

Mark explained that he worked as a farrier, shoeing horses, and had observed that the horses that had a spiral hair whorl above the eyes were more difficult to handle. His ideas sounded interesting, so we discussed how we could scientifically validate his observation. I told him that cattle have the same hair whorl patterns and that it would be easy to watch herds of cattle going through the squeeze chute for their vaccination and record the hair whorl position and the behavior while each animal was restrained in the chute.

The data we collected and published after observing more than 1,000 cattle clearly proved that animals that struggled during restraint were significantly more likely to have a hair whorl above the eyes. Since this study was in the early 1990s, many breeders have selected for more docile animals. This has resulted in fewer cattle with a hair whorl high above the eyes in the middle of the forehead. Today, in many herds, the highest hair whorl position will be slightly above the eyes. Cattle with hair whorls high on the forehead were more fearful and became agitated more easily in the squeeze chute.

The same principle applies to horses. Horses with a high hair whorl above the eyes or multiple whorls are usually more excitable and become fearful more easily. These horses are more likely to be traumatized and ruined by rough training methods than are horses with calmer genetic traits.

We further reported a relationship between quality of sperm morphology and hair whorl shape in yearling Angus bulls. Bulls with abnormal whorls — lines that usually run vertically between or just below the eyes, rather than spirals — were found to have more defective sperm than bulls with centrally located, normal-looking spiral whorls.

LOW HAIR WHORL: CALM

HIGH HAIR WHORL: FLIGHTY

ABNORMAL HAIR WHORL

EVALUATING SHEEP

Temperament tests that work well for British or European cattle breeds and bison may work poorly for sheep. Among the most passive of animals, sheep held tightly in a sheep restrainer may freeze instead of fighting restraint the way cattle do. This is called **tonic immobility**. (Purebred *Bos indicus* Brahman or Zebu cattle may also freeze; see page 42.)

Sheep's tendency to flock tightly is related to temperament. The calmer, less easily stressed sheep will flock more loosely.

Cattle, pigs, and goats will vocalize (bellow, squeal, or bleat) when they are frightened or hurt during handling and restraint. Sheep will remain silent when they are in pain because they are the most defenseless of the prey species. A lone lamb will vocalize when separated from the flock. This is due to separation stress, not fear or pain.

TEMPERAMENT, SAFETY, AND MEAT QUALITY

It is important for producers intending to deliver high-quality meat to their buyers to select for temperament. In auctions and slaughter plants, cattle that become highly agitated are dangerous for people to handle and accidents will happen.

Agitated animals are more likely than calmer animals to have dark cutting meat (meat that does not bloom or brighten when cut and exposed to air). Dark cutting meat is darker and drier than normal meat, and it has a shorter shelf life in the refrigerated meat case. Long-term stress, such as that experienced during a 15-hour truck ride, depletes the energy stores in the meat and may cause it to darken.

The incidence of dark cutting meat in the United States has more than doubled since the 1960s. This is due partly to genetic lines of cattle with excitable temperaments. Selection for large, lean muscles in pigs has resulted in tougher pork, and many breeders are switching to pig breeds with slightly slower growth rates to improve pork quality.

A study by Lily Edwards-Callaway of Colorado State University showed that calm handling during the last few minutes before slaughter is essential. Jamming in the chute or electric prod use increased pale, soft poor-quality pork. Our research shows that cattle that became highly agitated in the squeeze chute had lower weight gains and were more likely to be dark cutters.

It's worth reiterating that selecting for calm temperament is important, but overselecting for the calmest animals may have undesirable effects. The calmest animals might lose some other valuable trait, such as foraging or mothering ability. The emphasis should be on culling the highly excitable individuals instead of selecting for the absolute calmest.

The Dangers of Overselection for Single Traits

Genetic traits can be linked in unexpected ways. Single-trait selection for a behavioral trait can have unexpected consequences on body type and behavior.

Russian scientist Dmitry Belyaev studied the long-term negative effects of selecting for calm behavior in silver foxes. In roughly 10 years and 20 generations, selection of the calmest individuals resulted in heavy-boned foxes that resembled black-and-white Border Collies. Although the once-fierce foxes were indeed calm and friendly toward people, after many generations negative traits appeared: epilepsy, defective bending of the neck at birth, and sometimes eating their babies.

Grazing animals typically do not have the severe behavioral problems found in some intensively selected pigs and chickens, and the selection criteria are less intense for cattle and sheep than they are for pigs and chickens. Some high-producing egg layers and lean hybrid pigs have abnormally high excitable temperaments. This has led to cannibalism in chickens and increased tail biting in pigs. As evidenced by these animals and the fox experiment, a basic principle of genetic selection emerges: **Overselection for any single trait ruins the animal.**

RUSSIAN SILVER FOX

Bred to be gentle, this fox has a white tail tip. Some of the foxes resemble black-and-white Border Collies.

TEMPERAMENT MAY INVOLVE MORE THAN ONE TRAIT

A lamb will vocalize loudly when separated from his flock but be silent when he is injured. Defenseless prey animals such as sheep do not want to advertise to predators that they are in pain. Cattle and pigs will vocalize when hurt or scared during handling, but sheep do not. Cows bred for high maternal traits may be more motivated to defend their calves. All these traits are influenced by genetics.

The overall temperament of any grazing animal may consist of several different traits. As mentioned, in most animals, the traits for vigilance and fearfulness are linked. Often, high vigilance, high fear, and a fine-boned lightweight frame all go together. If a producer overselects exclusively for low fear, he may have a dull animal with reduced motivation to defend her calf.

An extensive ranch requires animals that are vigilant, can forage on rough terrain, and can defend their young. In contrast, when animals of any species are intensively managed and raised indoors or on closely managed pastures, the calmest (low-fear) animals will have higher weight gain and better conception rates after artificial insemination. Therefore, the animal that is best suited for an extensive environment is not the same one that is best suited for more intensive operations.

Unique selection criteria may disconnect the traits of vigilance, fear, and light bone structure. When the temperament tests in this book are conducted to an extreme, the result may be a nonreactive animal with a low hair whorl. For animals

raised under extensive range conditions, the tests should be used to eliminate "berserk" or "crazy" cattle and not be used to select for absolute calmest.

One final word of warning: There will be a point where the "gentleness" and "fights predators" traits will be at an optimal level. Continued overselection will result in some unpredictable bad consequences.

The Male of the Species

When bulls, rams, bucks, and boars grow up in a herd, with their mother present, they develop a healthier sense of identity than if they are bottle-fed or pampered by a human. This means that ultimately they will be safer for humans to handle, regardless of species.

UNDERSTANDING BULL BEHAVIOR

Bulls are considered to be the most dangerous of domestic animals. Their size and unpredictability alone should make handlers err on the side of caution. Bulls make up only 2 percent of the cattle population but are the number one cause of fatalities among humans who handle livestock. It's very important to understand their behavior to ensure safety.

The bull most likely to attack people is a bull calf that has been raised on the bottle by a person. That isolation from other cattle leads to a calf's mistaken identity. He sees himself as a human, and therefore, when he matures at 18 to 24 months, he will challenge people in efforts to be the "boss" of the herd. Orphan bull calves that cannot be raised with other cattle should be castrated to prevent future aggression problems. They must be castrated immediately, before the male hormones stimulate natural male behavior.

The best way to help prevent dangerous bull behavior is to rear young bulls on a mother cow in a social group with other cattle. Bull calves reared in this manner are the least aggressive because they identify with their own kind. The interaction with other cattle is needed to establish identity and behavior.

Fear Not the Same as Aggression

Fear and aggression are two behaviors not to be confused in bulls. A bull that becomes agitated in a squeeze chute is fearful, but a bull that chases or challenges a person on a pasture is aggressive and dangerous. That bull should go directly to a processing plant or, if valuable, to a secure bull stud facility.

Producers cull the really wild animals that break gates, charge people, and go berserk during handling. A single agitated animal can spread fear through the herd.

A Bull's Warning Signs

Bulls are the most dangerous of domestic animals. Although a bull will assume a threat stance prior to attacking, he usually will not look at his victim, and thus many people fail to recognize the warning signs. Here is how a bull prepares to charge someone he perceives as a potential rival.

1. The bull's threat stance begins with a broadside display — turning sideways and flexing his neck to show how big he is.

2. He turns to face the "threat" (the person) head-on, usually pawing and horning the ground.

3. Then he charges.

A bull performs a broadside threat before he attacks.

The emotions of fear and aggression are totally different and work in different brain systems. Frightened bulls can be calmed down and handled by people who are gentle with them. But just being nice to an aggressive bull that is imprinted to people cannot stop him from fighting you to be dominant. It is not a tameness issue; rather, aggression is the result of mistaken identity.

Interacting Safely with Bulls

Some herds of cattle are very tame, and the bull likes to be stroked and scratched. If he is well socialized with his own kind,

Herd animals may have a complicated social life. When these bulls stop fighting, the subordinate animal has room to run away. Fighting is often more severe in small pens because there is not room for the subordinate to get away.

you can do this safely. Stroke him under the chin or on the back and withers.

He must respect your space. If he pushes against you, instantly stop scratching — take away the reward. He will learn that he must stay outside your space if he wants to be scratched or fed a treat.

Never play butting games with calves. When they grow up, you risk serious injury, even when it's unintentional, because of the sheer size and power of the animal. Avoid scratching the forehead, because this may trigger the butting instinct.

People often wonder why a group of bulls that were compatible and got along out in the pasture will fight when confined together in a pen. This happens because the pen is too small for each male to have a well-defined territory around him. When moving bulls, be careful when pressuring their flight zone. They may attack if pressured too hard.

RAMS, BUCKS, AND OTHER MALES

The same principle — that the hand-reared male is most dangerous — applies to ram lambs, goats, deer, bison, and other grazing animals. Hand-reared male llamas may attack people when they become mature. A common name for this is berserk llama syndrome.

There have been two cases of hand-reared buck deer killing the person who raised them. When a male deer is fully mature and his antlers have shed their velvet coating, he becomes more aggressive when he goes into rut during mating season. In both cases, the pet buck deer misinterpreted his owner's actions as a challenge. In the first case the person knelt down to take a photo, and in the second case the person bent over.

The deer interpreted their actions as lowering the antlers for a challenge. In all grazing species, early castration of orphan males will prevent dangerous aggression. Steers and female grazing animals can be safely hand-reared.

Life within a Herd

Herd animals have a social life that may be invisible until you make time to simply watch them for a while, every day, through the seasons. Look to see which animals prefer to be together, which ones have higher status, which (if any) are isolated (a sign that something might be wrong), and which is the true leader when the herd moves. Sisters, moms, and daughters will often graze together. When ranchers give sequentially numbered ear tags, the following year they often observe that many cattle go through the chute in the same order.

This observation time will provide valuable insights to help you understand and work with your animals.

DOMINANCE HIERARCHIES

All species of grazing animals form dominance hierarchies. The biggest, strongest, highest-ranking animals will push away other animals from a water or feed trough. The top-ranking bull will chase smaller, subordinate males away from the female he wants to breed.

It's important to note that the true leader of the herd or flock is usually not the dominant animal.

When animals are driven or are moving by themselves to pasture or water, the individual that is dominant at the feed trough will be in the middle of the group, where he or she is safest from predators. The lead animal is usually a calm, confident animal. The flightiest animals are often in the back of the group.

A ranking system reduces conflict over mates and resources, although in some species, such as horses, lower-status animals continually challenge higher ones.

Dominant animals are more likely to be the larger individuals. Horned animals usually rank higher than hornless ones. Temperament also plays a part. Angus cattle are more aggressive and feisty than some other breeds, and a small aggressive

Don't Turn Your Back on a Dairy Bull

Dairy bulls tend to be much more aggressive than beef bulls because dairy calves are raised on their own rather than among their kin. The calves become imprinted on people and think they are people. Then, when they become sexually mature, instead of exerting dominance toward other bulls, they are assertive around and toward people.

When a bull shows signs of being threatening, don't turn and run. The safest thing to do is to look away and slowly back away. *Never* turn your back on a bull. Strongly consider culling any bull that threatens people.

This pig risks cutting its foot on the metal strip that goes across the top of the waterer. The pig is jumping up because the trough is too high.

Angus may win fights with a larger, more placid Hereford.

Dominance orders also have an effect on handling. Dominant animals usually travel in the middle of the herd, where they are safe from predators. When the herd gets highly fearful and mills around, the dominant animals will be in the middle of the bunch and the weakest animals will be pushed to the outer edges.

An intact male is more likely to attack a handler if he smells the scent of a lower-ranking animal on the person's clothes or hands. In situations where handlers are not protected by a chute, the dominant bull, ram, or boar should always be handled and touched first.

Herd hierarchies play out differently in different species, both domestic and wild.

Dairy cattle. Dominance hierarchies become most obvious at feeding time. High-producing cows need to be able to eat without being butted away from the feed. Low-ranking animals may lose body condition if they are continually pushed out. Watch carefully to ensure that subordinate animals are able to eat. In feeding systems where animals put their heads either in a locking stanchion or through vertical bars, it is often advisable to have a few extra spaces so that a subordinate is not forced to eat next to dominant animals that she fears.

Beef cattle. When feed is provided ad lib, where the cattle can eat all they want, the animals can learn to take turns eating. Keep hay racks and troughs full to ensure that low-ranking animals get feed. If a pelleted feed or hay is limited, it *must* be spread out in a line on the ground or fed in a long trough with sufficient space so that all animals have easy access.

Sheep. The same feeding principles apply to sheep. If sheep are pastured with cattle, steps must be taken to ensure that cattle do not shove sheep away from the feeder. There are some differences in how cattle and sheep fight. Beef and dairy bulls will fight by butting and shoving. Rams will back away and then make a running charge at the other animal. This is an instinctual hard-wired difference in the two species.

Pigs. There is a species difference in the fighting pattern among pigs. The dominant animal will bite and shove at the side of the lower-ranking pig's neck or shoulder. To help prevent fighting over feed, build small, solid partitions on feed troughs to keep pigs from seeing the neck or shoulders of the pig on each side of them. The partitions must be just long enough to cover both neck and shoulders.

Goats. Nanny goats can be very aggressive around the feeder and forcefully butt other goats away. When feed is limited, space them out. A single very aggressive animal may need to be fed separately.

All species. Observe carefully to ensure that low-ranking subordinate animals are maintaining their body condition and are not getting skinny. If a low-ranking animal starts to lose body condition, you may need to feed it in a separate area. Another approach is to take all bullies and pen them together.

MIXING STRANGE ANIMALS

Always mix strange animals together in a new pen that is neutral territory. Never introduce a strange animal into another animal's home pen. The resident may perceive it as an intruder invading its turf. Sometimes it is best to introduce new animals by letting them touch noses through a fence.

When the animals are first mixed you must carefully observe them. Some butting and scuffling is normal, but if one animal gets really aggressive, it must be removed. Normally, strangers will have some mild fights to determine a new dominance hierarchy. When the animals determine who is boss, the fighting should stop. The reason animals develop dominance hierarchies is so that herd members do not have to fight constantly.

Animals Reared Alone

Horses, cattle, dogs, and other animals that are reared alone never learn the give-and-take of social interactions with other animals. This may result in an adult animal that will constantly attack other animals. It has not learned that even though it has won a fight, it does not need to continue fighting to be dominant.

Many serious behavior problems can be prevented by raising young animals in social groups with other animals. Adult animals that have been reared alone in a small pen from a young age may have great difficulty integrating into living with groups of animals on a pasture or in a pen. Even when provided with lots of space, the animal that has been reared alone may continue to attack other animals. Some stallions are overly aggressive due to being raised alone without having the chance to learn from older animals, and to always being kept in a stall.

PART II

HANDLING LIVESTOCK HUMANELY AND SAFELY

3 | WORKING IN PASTURES AND PENS

Animal behavior on rangeland, in pastures, and in corrals is governed both by instinct and by learned responses to surroundings. Regardless of their genetic makeup, all species of grazing animals are born with natural behavior patterns. Over the eons, these patterns developed in wild and domestic herds as strategies to avoid predators.

For example, they will graze in bunches that make it harder for a predator to single out and target an individual for its next meal. Or they will all turn to face a predator, real or perceived, that is outside of their flight zone in any direction. Early naturalists referred to these hardwired predator-avoidance behaviors as instincts; modern animal behaviorists call them **fixed action patterns**. Skilled, quiet handlers can make use of these innate behaviors to gather and move livestock successfully.

When low-stress gathering and herding methods are first used on animals that are not accustomed to people, they trigger a progression of behavior patterns. These begin as purely reactive and instinctual and ultimately become based on animal learning and trust.

PRINCIPLES

Natural behavior patterns that are innate in all grazing animals enable them to avoid predators.

Stock handlers must understand the principles of the flight zone and the point of balance.

Use the principle of pressure and release. Find the right amount of pressure to keep the herd moving forward without causing the animals to run.

When cattle are moving where you want them to go, back off and relieve your pressure.

Stress will be reduced if you carefully introduce animals to corrals, restraint devices, and milking facilities before doing procedures.

When animals in a group are moved through a pasture gate, they should walk past the stockperson in a calm, controlled manner.

Males of different species exhibit the flehmen response, a hard-wired instinctual behavior responding to the presence of a female.

A sow learns by experience how to build the best nest for her babies.

Understanding Herd Instincts

Some instinctual behavior patterns are very rigid and resistant to change, whereas others can be modified by experience and learning. For example, a sexually mature bull will always respond to a cow in estrus by curling his upper lip, known as the **flehmen response**. This is an example of a fixed action pattern, performed the same way every time, regardless of experience.

A sow that lives in the woods or in a barn will build her babies a nest of leaves or twigs or whatever material is available. Farmers have observed that by the time she has her second or third litter of piglets, she begins to select and place her nesting materials more deliberately: the leaves may be organized on the inside of the nest, and twigs on the outside. This is an example of an instinct that can be modified by learning. A sow knows by instinct how to build a basic nest, but her nests improve with experience.

HERD OR FLOCK FORMATION

Living in groups rather than having a solitary existence improves the chance of survival for grazing animals. While some animals graze, others are on the lookout; any animal that sees any potential danger can alert the entire herd. Antelope flash the white under their tails to signal others of the presence of predators. Cattle and sheep will alert each other with head-up movements. Group living ensures constant vigilance, the safety that comes from large numbers.

When native herds of bison freely roamed the plains, they grazed one area of pasture and then moved on. Because there were lots of predators around, the animals grazed in tight bunches for safety, thereby "mowing" the grasses. After the grass was mowed, they moved on to a new area. Modern rotation, or mob, grazing systems mimic natural grazing behavior by densely stocking animals on a section of ground for a short period of time and then moving them to fresh pasture. This pattern improves the pasture because the herd's manure fertilizes the grass. The tight bunches of animals evenly mow an area instead of selectively picking only the most palatable plants.

Pasture specialist Fred Provenza says using mob grazing prevents animals from "eating the best and leaving the rest," and adds that the grass grows again when the animals are driven to a new area. Producers who use rotational grazing must allow the grazed pastures sufficient time to fully regrow. A common mistake is not allowing pastures that have been mob-grazed sufficient recovery time. There are huge variations in climate and in the time required for plants to regrow and be ready to be grazed again.

Defense Strategies of Grazing Animals

1. Turning and facing the predator when it enters the pressure zone. Another name for the pressure zone is zone of awareness.

2. Turning and moving away from the predator when it enters the flight zone

3. Running in the opposite direction when the predator crosses the point of balance

4. Loose bunching of the herd

5. Milling and circling of the herd

PRESSURE ZONE AND FLIGHT ZONE

Grazing animals naturally employ five basic instinctual behavior patterns to avoid predators. A handler who intuitively understands these behaviors should be able to gather and drive almost any grazing animal herd.

1. When grazing animals first spot a predator, they will turn and face it. The predator is in the pressure zone. The area in which an animal first becomes aware of a potential threat — whether a predator or an approaching handler — and turns to face it (or sometimes just turns its head) is called the pressure zone. The animal monitors the location of the threat in relation to himself and makes a decision about when it's safe to stay and when it's time to move away. Maintaining a certain distance from the intruder gives an individual a head start if pursued. The turning-and-facing behavior is hardwired, but the flight distance is affected by experience. According to Ron Gill at Texas A&M University, animals that are being moved by a handler want to see where that handler is. Curt Pate, a cattle handling specialist, uses this tendency as a method to "draw" cattle toward him when he wants to sort them. Calm cattle that are paying attention to a handler will tend to follow when the handler backs away. Curt makes sure the desired animal is looking at him when he starts backing away.

Animals are very sensitive to body posture. They know the difference between a stalking lion and one that is simply walking by, and they can read human posture and intent just as well. When grazing animals learn to trust a calm and respectful handler, their tendency to turn and face the person in the pressure zone will decrease, and instead they will become more willing to walk away in a straight line. The natural turn-and-face behavior has been overridden by learning.

When the animals become aware of the predator's or handler's approach they will turn and face him or her. This indicates that the predator or handler has entered the **pressure zone**.

Experts at Reading Body Language

Grazing animals have evolved to be highly sensitive to body posture and intention. For instance, when lions are not hunting, antelope will follow them at a distance they perceive to be safe. If a lion begins to exhibit stalking behavior, the safe distance expands, the flight zone grows larger, and the antelope flee.

PRESSURE ZONE.
The single tan animal has lifted its head and is aware of the handler, who has entered the outer edge of the pressure zone.

NO REACTION.
The animals in the distance are unaware of the handler's presence.

FLIGHT ZONE.
The four black cattle have turned away because the handler walked through the pressure zone and entered the edge of the flight zone.

When the handler moves closer and penetrates the flight zone, the animals will turn and move away. This image shows three stages of the animals' reaction to an approaching person.

2. At the point where the animals can no longer tolerate the handler's approach, they will turn and move away. The handler has entered the flight zone. As the handler approaches closer to the animal, he leaves the pressure zone and enters the flight zone, and the animal will turn and move away.

3. If a handler crosses a grazing animal's point of balance, located at the shoulder or just behind the eye, the animal will always run in the opposite direction. This hardwired, innate maneuver helps an animal dodge a fatal attack on its flank.

A human handler can take advantage of this instinctual response, by passing across the **point of balance** to move livestock calmly. Using point-of-balance principles is especially helpful when guiding either a single animal or a group of animals through a single-file or double-file chute. Handlers who want to move an animal forward must never stand at its head and poke it on the rear. This gives the animal conflicting directional signals. When you work inside the flight zone and walk in the opposite direction of the desired movement, the animal will move forward when you cross the point of balance.

4. Grazing animals form bunches when they live in an area with predators. This makes it harder for a predator to single out a lone individual.

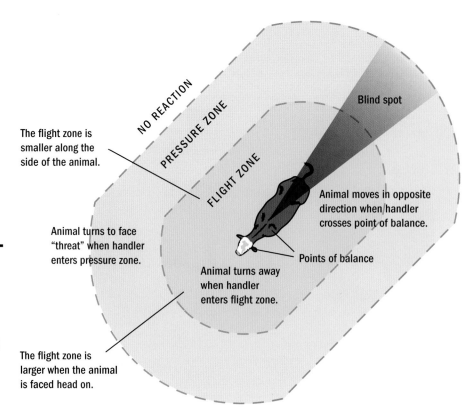

NO REACTION

PRESSURE ZONE

FLIGHT ZONE

Blind spot

The flight zone is smaller along the side of the animal.

Animal moves in opposite direction when handler crosses point of balance.

Animal turns to face "threat" when handler enters pressure zone.

Points of balance

Animal turns away when handler enters flight zone.

The point of balance is located at the shoulder when the handler is very close to the animal, and it moves to just behind the eye when the handler is farther away.

The flight zone is larger when the animal is faced head on.

Cattle in bear country will naturally graze in loose bunches. Those in areas generally free of predators tend to spread out. Livestock will remain calm when grazing in loose bunches. Cattle herding specialist Bob Kinford states that when cattle become acclimated to grazing in bunches they will all face in the same direction as they mow the grass.

Dances with Cattle

Dawn Hnatow, livestock manager at Cattle Up Ranch in Sulphur Springs, Texas, explains that it is important to teach cattle to move in a controlled way. A skilled stockperson initiates cattle movement by entering the flight zone, and she stops it by retreating back into the pressure zone. She relieves the pressure when the animals cooperate. This teaches the cattle that the handler will "ask" them to move, and once they have moved, the uncomfortable penetration of the flight zone is relieved.

Flighty breeds of sheep will form a tight flock when threatened. Flocking is their main defense, and they can group together very quickly. I have observed sheep forming a tight bunch almost instantly upon spotting the shepherd's herding dog.

5. When predators attack livestock, the flock or herd begins milling and circling. Dominant animals move to the middle of the tight circle (the safest area), and the weakest ones pace and mill at the outer edges of the circle.

Instinctive fear-motivated behavior, milling and circling can be an effective defense when predators attack, especially effective with large herds. The adult females face outward and attack predators with their horns. The predators eat a few animals on the perimeter of the milling mob, and the rest survive. Ranchers who graze animals in country with lots of predators have reported that animals that scatter when attacked will lose more lambs or calves than animals that stand their ground by bunching together.

Milling is stimulated by fear, but standing one's ground is more likely to be motivated by a mother animal's instinct to protect her young. A mother with a calf will often face a predator and may attempt to attack it. There are genetic influences on defense behaviors, and some animals are more successful in defending themselves from predators than others. In South and East Africa, a cavalry of Cape buffalo moms will band together and chase lions away from a threatened calf.

Milling cattle are frightened and highly stressed. People handling cattle must avoid triggering this predator-avoidant milling behavior. It is a hardwired behavior that good stockpersons never want to see when animals are being moved or handled.

The dominant animals will be in the center of these milling cattle. A good stockperson never wants to see milling cattle. It is a sign of extreme fear and stress. Avoid situations that lead to milling behavior.

Bud's Basics

Cattle handling specialist Bud Williams developed a basic principle. To speed up forward movement, he moves within the flight zone in the opposite direction of desired movement; to slow down movement, Bud moves outside the flight zone in the same direction as desired movement. (He remains within the pressure zone so that the animal will remain aware of his presence.)

Handling Practices That Use Animal Instinct

Once you understand the instinctive patterns that determine how an individual animal or herd will move in response to specific kinds of interaction with other animals and people, you can capitalize on that knowledge. Getting inside the skin of the animal to think the way he thinks, so to speak, and recognizing his response to your own movements and appearance when you are handling him will enable you to gather, herd, and turn animals in all kinds of situations.

TAKING ADVANTAGE OF NATURAL FOLLOWING BEHAVIOR

Single-file following is an instinctive behavior pattern some grazing animals use when they walk to water or walk in from grazing to bed down at night. They are calm when they string out and are not worried about predators. You can use this behavior to help move animals down an alley and through a chute. When a cow or pig sees another cow or pig up ahead, he will follow. Note that sheep are more inclined to stick together, as they are more defenseless.

STARTING, STOPPING, AND TURNING ANIMALS

A skilled stockperson can move calm cattle with just one or two steps. With practice, a handler can learn to stop and start a single animal by moving a step backward or a step forward. It is strongly recommended for people who are first learning low-stress principles to practice starting and stopping cattle. (This will not work with completely tame cattle with no flight zone. Tame cattle should be led, because they do not respond to pressure on the flight zone.)

Calm cattle have a natural instinct to turn and look where the handler is and, compared with agitated cattle, will be more inclined to always keep the handler in view. Cattle that have had experience with a quiet handler who relieves pressure on the flight zone when they are moving in the desired direction will often stop constantly turning to see the handler. Steve Cote, author of *Stockmanship: A Powerful Tool for Grazing Lands Management*, explains that when cattle learn to trust the handler they stop turning around to look while being driven. They will learn to drive straight, and the learning will override the turning instinct.

Nancy Irlbeck from Wellington, Colorado, demonstrates how her sheep quietly move around the "bubble" that is formed by the outer edge of the flight zone.

Using Loose Bunching Behavior to Gather Animals

To calmly gather livestock from large pastures or pens, you walk slowly back and forth on the outer edge of the herd's collective pressure zone. The herd is aware of your presence, but you stay far enough back to avoid penetrating the flight zone and causing the herd to move.

This triggers the innate protective instinct in the animals' brains to form a loose bunch. The loose bunch should be formed before you attempt to enter the flight zone to move the group. Premature entry into the flight zone can cause animals to scatter.

When you are moving a bunched group of cattle to a new location, all of the cattle should be heading in the same direction and walking quietly. They should not be bumping into each other or turning. If they start doing this, it indicates that the animals are becoming agitated and planning to flee. Walk slowly and take great care never to cause the cattle to start milling or running.

Coaxing cattle to bunch by "stalking" them on the edge of the collective pressure zone is probably slightly stressful for them

until they get used to it. But when you handle them in this quiet manner on a regular basis, they soon recognize that you are not going to apply too much pressure and invade the flight zone too deeply.

With practice, cattle learn that your pressure on the outer edge of the collective flight zone is relieved when they move in the desired direction.

UNDERSTAND THE COLLECTIVE FLIGHT ZONE

Both the collective flight and pressure zones for a herd of grazing animals vary according to the size of the herd, the amount of contact with people, and genetic factors influencing temperament.

Animals naturally follow the leader. Take advantage of this instinct to facilitate calm, quiet animal movement.

Differences in Goat Behavior

Goats have some differences in their behavior compared with cattle or sheep. They do not flow as easily through a single-file chute, and they are more likely to lie down and refuse to move (a behavior known as **sulking**). When goats are moved in a group, the older females (**queens**) will proceed first. Handlers must be careful not to overcrowd goats in the corner of a corral, because the animals may pile up and become injured.

Working the Flight Zone

The actual distance at which an animal moves away from another animal or a human varies among species and even among breeds within a species. Some animal groups and individuals tolerate a very close proximity, whereas others will move away when anyone comes within 20 feet (6 m). These individual and breed differences exist for a number of reasons:

- Unique genetics determine the degree of reactivity or fearfulness an animal will exhibit when suddenly confronted, accounting for why some individuals within a breed have larger flight zones than others.

- Experience with calm, respectful stockpersons and handling procedures will reduce flight distance; rough handling procedures will increase flight distance.

- Animals will have a larger flight distance from a person they associate with a rough handling experience than they will from a person who has handled them gently.

- Frequent contact with people will usually result in an animal developing a shorter flight distance. Animals that seldom see people will have less tolerance and a larger flight distance.

- The angle of a person's approach also accounts for differences in flight zone. Approaching animals head-on can increase flight distance.

- A solid barrier placed between the person and the animals will decrease the flight zone.

Flight-zone size is related to the tameness or wildness (fearfulness) of the animal. These sheep are circling around the handlers and maintaining their flight distance. The handlers have a "bubble" like a force field around them. Handlers can use and direct this bubble. If the animal has no flight zone and is completely tame, this will not work.

Working the edge of the flight zone can make moving livestock easy, using the principle of pressure and release. If cows on range or pasture or in a pen turn and look at you, you have entered the outer edge of the pressure zone, outside of the collective flight zone. Being outside the flight zone makes them turn toward you so they can determine if you are a potential threat. When the handler penetrates the outer edge of the flight zone, the animals will turn away. Cattle, sheep, and pigs respond in a calmer manner to flight-zone penetration than do bison. Bison tend to break into a run when their flight zone is penetrated.

Keep in mind that you must be close enough to make the animals move but not so close as to cause them to panic and flee. If cattle start moving too fast, back off and get out of the flight zone. When they slow, move back into the outer edge of the flight zone to keep them moving. *Never* exert continuous pressure on the flight zone.

Animals confined in a single-file chute or other confined area where they are not able to move away may become agitated or rear when a person stands deep in their flight zone. An animal will often calm down and stop rearing when the handler backs away, out of its flight zone. Never attempt to push a rearing animal down. That may make him more agitated because he wants to get away from you.

Working the Point of Balance

To move a single animal, stand behind the point of balance and stay out of the blind spot directly behind the animal. (In later sections of this chapter, you will learn to move groups.) When you are close to the animal, the point of balance will be at the shoulder. When you are farther away, the point of balance may move forward, to just behind the eye. When you are on the outer edge of the pressure zone, the animal becomes aware of your presence and turns around to face you. When you penetrate the edge of the flight zone, the animal turns away and moves.

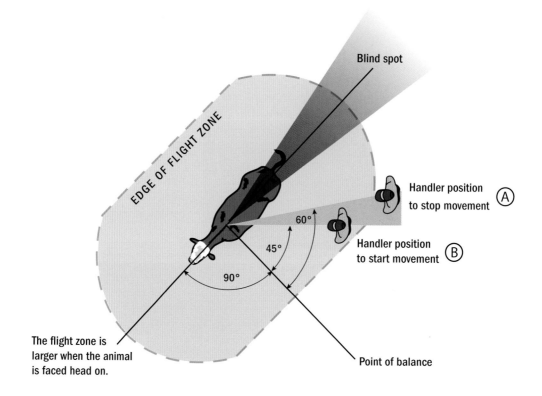

Blind spot

EDGE OF FLIGHT ZONE

Handler position to stop movement Ⓐ

60°

Handler position to start movement Ⓑ

45°

90°

The flight zone is larger when the animal is faced head on.

Point of balance

To move a single grazing animal forward, you must be behind the point of balance. Moving in front of the point of balance at the shoulder will make the animal turn around and head in the other direction on a pasture or go backward in a single-file chute. When you want to initiate movement, approach just behind the point of balance and move from position A to position B. Positions A and B work really well for moving either a single animal or animals in a chute. When a group of animals is flowing calmly by you through a gate, do not keep trying to move back to positions A and B. Position yourself so that the animals flow around the "bubble" formed by their collective flight distances.

By recognizing a herd's collective flight zone, you can adjust your movements to suit the animals better. For example:

- Animals with genetics that cause them to be flighty will have a larger collective flight zone than animals with calm genetics.

- Herds that have experienced lots of quiet contact with people have a smaller collective flight zone.

- Extremely tame herds with little or no flight zone usually respond best to being led instead of being herded.

USE PRESSURE AND RELEASE

A handler enters the outer edges of the collective flight zone to make the animals move and retreats from the collective flight zone back into the pressure zone to prevent them from moving too quickly. A basic principle is this: When they are moving where you want them to go, back off and relieve pressure on the flight zone. To prevent a herd or flock from running, *never* apply continuous pressure to the collective flight zone.

You don't want the herd to move any faster than a walk or slow trot, and you should always move among the animals at a normal walking speed (or emulate that speed if you are on horseback or in a vehicle). Stay quiet, slow, and steady, and refrain from waving arms. Do not become impatient with slow animals. Move the herd at the same pace as the slowest cattle.

These principles will be much harder to implement with bison. Compared with cattle, bison have a much stronger tendency to run away when a flight zone is penetrated. They get frightened and highly agitated easily, and they attack more often. Producers should consider training bison to go through chutes and perform other movements with food rewards.

WORK THE COLLECTIVE POINT OF BALANCE

A grazing animal moving in a group will position himself just behind the point of balance at the shoulder of the animal in front of it. Working multiple points of balance of a group can be used to make a herd move. A person moving inside the collective flight zone *in the opposite direction* of desired movement will speed up herd movement. A person moving *in the same direction* just outside the collective flight zone slows herd movement.

Comparing Cattle and Sheep

The movement pattern used to loosely bunch cattle is a slow-motion version of the movements a good sheepdog uses to gather sheep and induce them to flock. It is much easier to trigger the flocking instinct in sheep than it is in cattle. Sheep have a hyper-bunching instinct. A few quick movements of the dog can induce sheep to bunch, but quick movements are likely to make cattle or bison scatter. The basic flocking behavior is the same for all grazing animals, but it is much more intense in defenseless animals such as sheep.

Turning an Animal

TURNING A SINGLE ANIMAL

When using the hip method for turning an animal (below left), the handler starts at the edge of the flight zone at the shoulder point of balance and enters the flight zone while walking at an angle back toward the hip. If the handler walks toward the left hip, the animal will turn right; if the handler walks toward the right hip, the animal will turn left.

You can also turn an animal by approaching his head (below right). The handler starts outside the flight zone, in a position parallel to the nose, and enters the flight zone approaching on a slight angle toward the animal's ear. A skilled stockperson can learn to stop and start an animal by taking a single step forward or backward. If cattle have had experience with a quiet handler, who relieves pressure on the flight zone when they move, they will not run away.

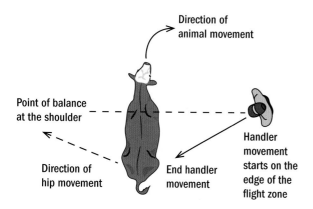

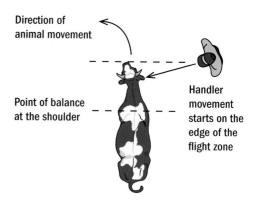

Source: Adapted from Steve Cote's cattle handling methods in *Stockmanship: A Powerful Tool for Grazing Lands Management* (USDA Natural Resources Conservation Service, 2003)

TURNING A SINGLE ANIMAL BY ENTERING HIS BLIND SPOT

These diagrams show how to turn a single animal by entering the animal's blind spot. When a handler enters the blind spot, the animal will turn his head toward the handler because it wants to see the potential threat. If the handler walks through the blind spot to the other side of the animal, the animal will turn his head in the opposite direction of the handler's movement. Curt Pate uses this principle of attracting the leader's eye to move and turn groups of cattle.

Adapted from Curt Pate

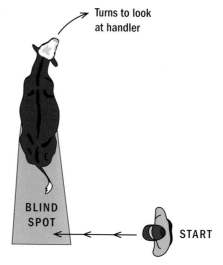

TURNING THE LEADER BY ATTRACTING ITS EYE

This diagram shows how to turn a small group of cattle by walking far to one side of the group to attract the leader's eye. This technique of attracting the leader's eye to change a group's movements works really well with purebred Brahman cattle, but they may insist on following you and be difficult to drive. Often it is easier to lead purebred Brahmans than it is to drive them. This is another illustration of genetic differences in innate instinctual cattle behaviors.

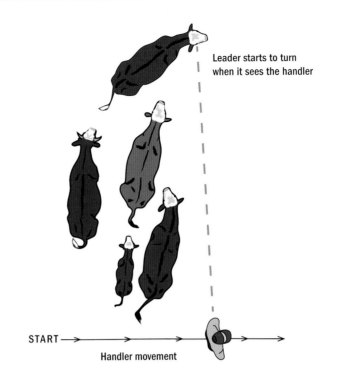

Leader starts to turn when it sees the handler

START →

Handler movement

CHANGING DIRECTION BY ATTRACTING THE LEADER'S OTHER EYE

Cattle handling specialist Curt Pate has developed a movement pattern that seems completely counterintuitive. The handler moves through the blind spot to the opposite side to attract the leader's eye. This movement will turn the head in the opposite direction from which it was traveling. It takes time to learn these patterns, and the handler has to spend some time training the herd to move calmly.

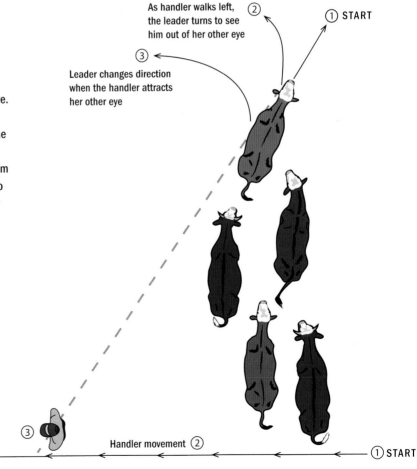

As handler walks left, the leader turns to see him out of her other eye ②

① START

③

Leader changes direction when the handler attracts her other eye

③

Handler movement ②

① START

STAY OUT OF BLIND SPOTS

Do not stand behind animals where they are unable to see you. Standing in the blind spot will hinder movement. Standing in the blind spot behind an animal's rear often causes the animal to turn and face you instead of moving where you want it to go. Never apply pressure to the flight zone from directly behind the animal in its blind spot.

APPLYING THE GATHERING AND BUNCHING PRINCIPLES

The majority of animals in the grazing herd *must* be loosely bunched before you enter the flight zone to move the entire herd. After animals that are closest to you turn and face you, you can achieve loose bunching by applying light pressure on the outer edge of the collective pressure zone. If you apply too much pressure before the cattle are loosely bunched, they will scatter.

Depending on herd size, previous experience, genetics, and terrain, it usually takes from 5 to 20 minutes to coax the herd to form a loose bunch. The right amount of light pressure just outside of the flight zone will make the animals group loosely, without making the leaders move forward. The handler needs to work inside the pressure zone so the animals are aware that he or she is there.

Don't circle too far around the herd. Your pattern of movement is similar to the movement of a car windshield wiper. Your "arc" should be no more than a quarter circle, depending on the size of the herd and the size of the space being used. Confined spaces require smaller arc movement, while large pastures require larger.

Some handlers prefer a straight zigzag pattern. In either practice, your movements must be *perpendicular* to the direction of desired movement, not in a circle around the herd.

Making 6 to 20 wide back-and-forth movements on the outer edge of the collective pressure zone will induce the herd to form a loose bunch. On ranches where cattle are spread out on rugged terrain, several loose bunches may form before the entire herd comes together.

Bud Williams's Genius

Gathering and loose-bunching principles developed by handling expert Bud Williams and used on cattle, farmed elk, and reindeer are reputed among producers to be highly, and miraculously, effective. They work so well, in fact, that when ranchers first observed the "cow whisperer" at work, they were sure he used magic to get them to move. I, too, thought he was a wizard when I saw him go into a pasture of strange cattle, gather them up, and bring them in.

After coming to understand animal behavior and instinct, I still think he had an amazing gift. The principles he developed can be employed by others on their pastures and rangelands when working their animals. His methods, however, may not work with extremely tame animals, which may respond best to being led.

THOUGHTS AND REFLECTIONS
Science vs. Intuition in Animal Handling

There are two schools of thought when working with animals. One is intuitive, and the other is based on diagramming and explaining animal behavior. There is a place for both approaches.

Most people who first start working with animals will learn more quickly when they have diagrams and clear explanations of behavior. One must remember, though, that a diagram cannot explain every situation and that your intuition and instincts play an essential role.

As your experience grows, you will learn how to influence the animal's movements by making very small changes in posture or position. This requires an attitude of quiet confidence. Animals are very sensitive to your emotions. They pick up subtle cues and know when you are angry and when you are calm.

An effective stockperson is never hesitant. When pressure needs to be applied to the flight zone, it is applied in a decisive, confident manner to give the animal a clear signal. When the animal responds, the handler retreats to reward the animal for going in the desired direction.

Some tips:

- Move back and forth continuously, because stopping and lingering too long in one animal's blind spot will make the cattle turn back and look at you.

- Make sure to move far enough to the side so that the lead animals can see you.

- You can give the rear animals a hard stare to get them to move. To do this, stand straight and tall, look right at an animal, and lock your eyes onto his. This further simulates the initial stalking behavior of a predator who is sizing up the herd.

- Individual cattle scattered off to one side of the pasture will be attracted as the herd moves into a loose bunch. Animals hidden in brush or timber will also be drawn out.

- On open pastures, it is important to take your time.

Never chase after stragglers or single cows. They will not want to be left alone and will eventually be attracted by and seek the safety of the moving herd. It is also important to resist the urge to press cattle into loose bunching too quickly, because they may scatter. The goal here is to cause only slight anxiety by simulating predator stalking behavior to make the animals want to bunch closely for safety. As you quietly work with your animals, their anxiety will be replaced by trust. The animals will learn that you will relieve pressure on the flight zone when they go in the direction you want.

WINDSHIELD-WIPER PATTERN TO INDUCE LOOSE BUNCHING

The handler moves back and forth on the outer edge of the pressure zone to induce the animals to form a loose bunch. The handler must zigzag without moving forward to induce loose bunching.

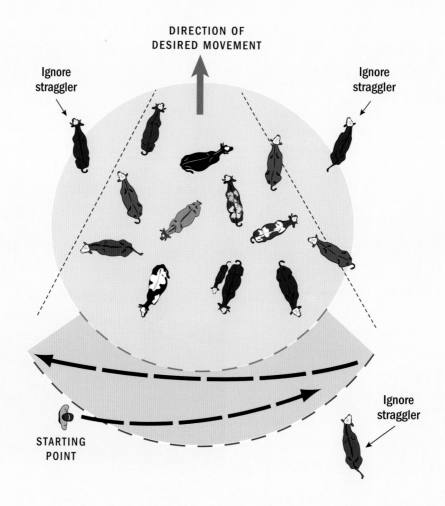

DIRECTION OF DESIRED MOVEMENT

Ignore straggler

Ignore straggler

Ignore straggler

STARTING POINT

THOUGHTS AND REFLECTIONS
Good Movement vs. Bad Movement

Every time you are working your animals, you are training them. You can train them to be easy to handle and have "good" movement, or you can train them to be difficult and have "bad" movement.

Good movement occurs when animals are all going in the desired direction at a calm walk. The herd should look like it is calmly walking toward water or making some other voluntary movement on a large pasture. Good movement resembles a flock of geese flying in a V-formation. The leaders are in front of the others, forming a loose V. In large groups, good movement entices the animals to follow but can be disrupted when the animals cannot see the position of the handler. All livestock species need to know where the handler is located.

Bad movement happens when animals are running, circling, milling, cutting back, or turning back to see the handler. The first signs of bad movement are stopping, wavering in the direction the animals are moving, and turning away to look at the handler. Bad movement in a few animals can prevent others from following in an orderly manner.

Initiating and Controlling Movement

The handler movement patterns presented in this section work because they trigger innate antipredator programs that are hardwired in the herd animal's brain. Learning to use the instinctual movement patterns of cattle is the next step.

Always induce loose bunching before attempting to move the entire herd forward. Once the majority of the animals have come together into a loose bunch, you can begin to move the herd or flock. To initiate movement, continue the back-and-forth arc and move forward to increase pressure on the collective flight zone. This will cause the herd to move forward and string out.

This is the time to use the principles of pressure and release. You need to find the right amount of pressure to keep the herd moving forward while avoiding too much pressure, which will cause the cattle to run.

The ABCs of Pressure and Release

A. Apply pressure to make cattle move; then back off and reduce pressure when they go where you want them to go. Never chase after a straggler.

B. Do not apply pressure when cattle are moving where you want them to go.

C. Develop a rhythm for repeatedly entering and moving out of the collective flight zone. Reenter the collective flight zone and reapply pressure when cattle slow down.

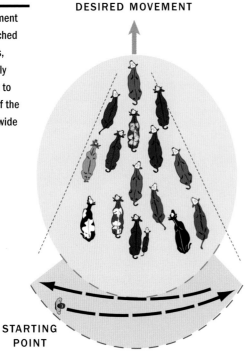

DIRECTION OF DESIRED MOVEMENT

To initiate movement in a loosely bunched group of animals, the handler lightly applies pressure to the outer edge of the flight zone, in a wide zigzag pattern.

STARTING POINT

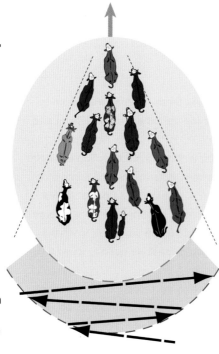

When the group has begun moving forward and is heading in the right direction, the span of the zigzag narrows. Practice the principles of pressure and release as you move the animals along. Be sure to reduce the pressure on the collective flight zone and retreat back into the outer edges of the pressure zone when the herd speeds up to lessen the incidence of running.

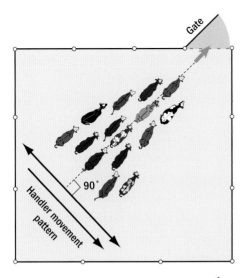

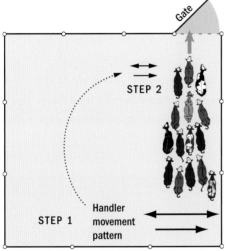

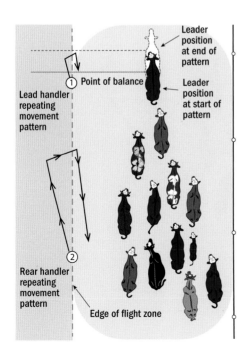

CONTROLLING DIRECTION

A herd of animals is like a car. Before you can steer, the car must be moving. When good movement is achieved with all the animals walking in the same direction, it is possible to control where they go. Put simply, the herd will move away from the direction of pressure you apply to the collective flight zone. This basic principle can be used for moving cattle out of large pens in feedlots or on ranches. The movement pattern should resemble a T-square with the pivot point on the exit gate.

TOP: DIAGRAM OF T-SQUARE PATTERN FOR A SINGLE HANDLER MOVING A HERD OUT OF A LARGE ENCLOSURE
The back-and-forth sweeping movements behind the group are made at a 90-degree angle to the direction of the desired movement. They are perpendicular to the animals' movements.

MIDDLE: MOVEMENT AROUND THE "BUBBLE"
Move to the gate to induce the animals to flow around you and go out of the gate. As the group of cattle approaches the gate, you must shift your position to head the cattle out of the gate. When you stand in the correct position, animals will flow around the "bubble" that is formed by the collective flight zone. It is like a force field around you.

BOTTOM: MOVEMENT PATTERN TO KEEP THE HERD MOVING ALONG A FENCE
The principle is to walk inside the flight zone in the opposite direction of animal movement to speed up movement and to walk outside the flight zone but just inside the pressure zone to slow the group down.

KEEPING THE HERD MOVING

The illustration below shows the triangle movement pattern to keep a herd moving in an orderly manner. This pattern works either along a fence or in open pasture.

If you are working alone: Use the Handler 2 movement pattern shown on page 83. As the herd moves, walk forward at an angle that gradually relieves pressure on the herd's collective flight zone. When the animals start to slow down, increase pressure on the flight zone by walking straight toward them. As they speed up, turn and walk back alongside and parallel to them in the opposite direction of their travel. As you cross the point of balance of each animal, he speeds up and moves forward. Walk at a slight angle toward the animals to increase pressure on the flight zone. Repeat the pattern continually to maintain movement.

It requires practice to determine the length of each movement pattern. If you walk along parallel to the herd, the herd will split. *Remember:* To quicken the pace of the herd, move inside the flight zone in the opposite direction of the desired movement. To slow the herd, move outside the flight zone but still within the pressure zone in the same direction of the desired movement.

If you are working with a partner: Bob Kinford warns that after the loose bunch has formed, the handlers should work the sides (see below) and front of the herd. Steadily pushing the herd from the rear is like tailgating another car and may increase stress.

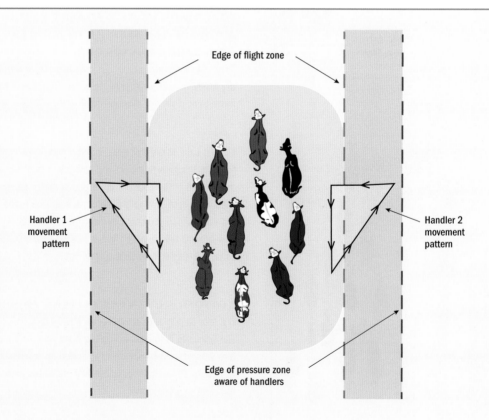

TRIANGLE MOVEMENT PATTERN
When cattle are being moved on open pasture with no fences, two people can keep the herd moving by using a triangular pattern on both sides of the herd. Keep the angles sharp, and do not fall into the habit of turning the triangles into circles. This diagram is adapted from Guy Glosson, Mesquite Grove Ranch, Jayton, Texas.

Edge of flight zone

Handler 1 movement pattern

Handler 2 movement pattern

Edge of pressure zone aware of handlers

NEVER CHASE STRAGGLERS

If one or more animals break away and straggle to the rear, resist the urge to get behind them and chase them. Allow the herd's or flock's movement to attract the stragglers.

At a walk, approach the stragglers at an angle that gradually increases pressure on their flight zone. Approach the animals just to one side of their heads and move just past the point of balance at the shoulder of the last straggler. As soon as the movement of the herd attracts the stragglers, start repeating the handler 2 pattern to keep the herd moving.

Be careful not to push the stragglers into the dominant animal in the middle of the herd, where these subordinates could be bunted or gored and become upset.

CONTROL ANIMAL MOVEMENT THROUGH GATES

Livestock movements should be under your control at all times. It's important for animals to understand that you're in charge. Never allow them to run wildly out of a corral or through a pasture gate and discover that they can escape from you. When that happens, it encourages bad movement.

To reinforce good movement, make them walk quietly past you at the gate as they go in or out of a corral or pen. If they have to walk by you, they will learn to enter in a more orderly manner. After the animals have passed through the gate but before you walk away, let all the animals turn back and look at you. If a group of cattle become agitated, give them a 30-minute break to calm down.

RETURNING STRAGGLERS TO THE GROUP

Rather than chasing stragglers, apply pressure on their flight zone in a manner that pushes them back toward the herd.

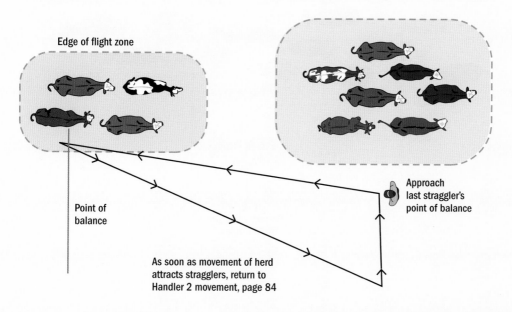

Edge of flight zone

Point of balance

As soon as movement of herd attracts stragglers, return to Handler 2 movement, page 84

Approach last straggler's point of balance

MOVEMENT IN AND OUT OF CORRALS AND PENS

The diagram below illustrates the correct position for the lead handler as a herd or flock enters a corral. Increase and decrease pressure on the flight zone by moving perpendicularly to the herd (and parallel to the corral gate) toward and away from the animals, repeatedly. Do not move parallel to the herd, back and forth alongside of the animals. You must apply enough pressure to keep them from veering away from the fence alongside them, but not so much as to cause panic.

Emptying Pens in a Controlled Manner

To empty the pen in a controlled manner, move in a small triangular pattern. Walk first to just inside the flight zone and then alongside the animals at the gate in the opposite direction of the flow. Then move in the same direction at a slight angle away from the animals, decreasing pressure on the flight zone. Finally, walk straight toward the animals and repeat the pattern as shown in the illustration on page 87.

To speed animals up, move across the point of balance inside the flight zone in the opposite direction of the desired movement. To slow animals down, move outside the flight zone in the same direction as the desired movement. To control the movement of the animals out of a gate, move to the sorting positions. Sometimes almost no movement is required and the animals will flow around a "bubble" formed by the flight zone.

ORDERLY MOVEMENT INTO A CORRAL

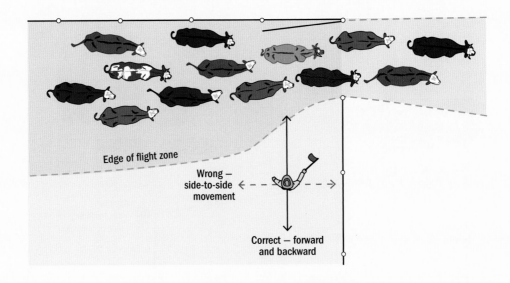

Edge of flight zone

Wrong — side-to-side movement

Correct — forward and backward

Sorting Cattle

To sort livestock, move forward and backward just inside and outside of the gate. Increase pressure on the flight zone of any animal you want to hold back, and decrease pressure on the flight zone of an animal you wish to let go by.

You can also work slightly farther in the pen and use Curt Pate's method: Instead of trying to hold back the animal you don't want, you attract the attention of the animal you do want to move out the gate. Curt calls this "hooking on." After the desired animal is "hooked on" to you, back up to bring it forward. Using this method, you can easily separate cows from calves.

Mimicking the predatory stare by locking your eyes on any cattle you want to hold back also works well.

Aussie Sorting Trick

When I visited the Australian outback, I learned a sorting trick that can be used in any set of corrals. Animals have a natural behavior to go back to where they came from. This instinctual behavior can be used to make sorting easier. When sorting from one pen into another pen, move all the animals into a new pen. Then sort them back through the same gate back into the pen they just left. Let the ones you want go by you into the old pen and hold the others back in the new pen.

EMPTYING PENS IN A CONTROLLED MANNER

The handler with the flag moves in a triangle pattern. When the handler is positioned correctly the cattle will flow out the gate around the "bubble" formed by the collective flight zone.

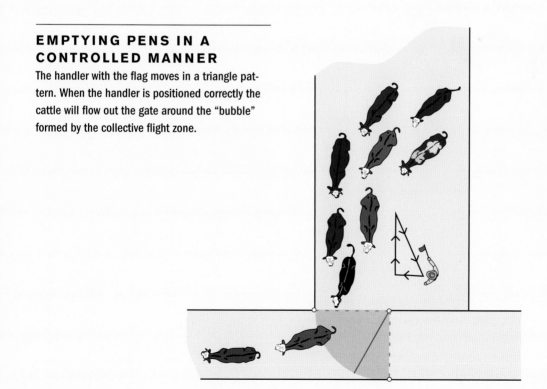

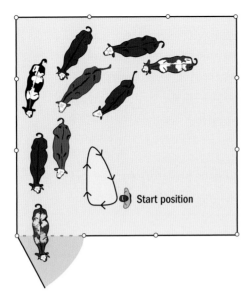

MOVING ANIMALS QUIETLY OUT THROUGH A GATE
Position yourself so that the animals will flow around the perimeter of the "bubble" formed by their flight zone. Imagine it is like a force field around you; the size of the force field is determined by the animal's flight zone or flight distance.

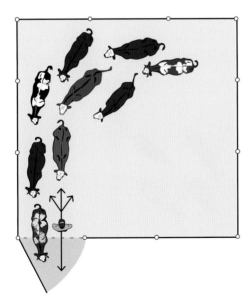

MOVEMENT PATTERN TO SORT ANIMALS FROM ALLEYS AND GATES
Handlers should work at the front of the group to control cattle movement. In calm animals, the point of balance will move forward; during sorting, focus on the animal's nose. Curt Pate explains that you should work the nose. The animal will go where the nose is pointed. Enter the flight zone when the nose is pointed to where you want the animal to go.

Body posture, too, can be used for sorting. A full-frontal posture will make the flight zone larger; presenting a narrow sideways posture will make the flight zone smaller. Turn sideways and present a narrow profile when you want an animal to go by you. Look down and away from animals you want out of the pen.

This method is very effective in both alleys and at gates, and it can be supplemented by holding a sorting stick with a flag on the end or a paddle. Keep in mind that blocking an animal's vision on one side of the paddle or flag will cause it to turn. (Note that the sorting stick is never to be used as a prod on the animal.)

Sorting Sheep and Brahman Cattle

Methods that work well for sorting English and Continental cattle breeds will work poorly on sheep that have a strong flocking and following instinct. These animals should be sorted in a single-file chute equipped with a sorting (drafting) gate. As the sheep move through the chute, they are sorted one at a time with gates.

Purebred Brahman cattle also may be difficult to sort in a large pen because of their intense following behavior. On many ranches, they are sorted through a single-file chute. Often the best method for moving purebred Brahmans is to train them to follow a person or horse.

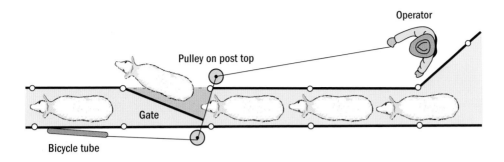

Operator

Pulley on post top

Gate

Bicycle tube

SHEPHERD'S TRICK

With this sheep-sorting system, one person can open and close the gate and at the same time drive the sheep. A bicycle inner tube serves as a big rubber band to pull the gate into an open position, so that you need only one rope to operate it. Adapted from Adrian Barber, South Australia Department of Agriculture.

Acclimate Animals to Handling and Milking Facilities to Reduce Stress

When you introduce cattle and other livestock to a new set of corrals or milking or handling facilities, avoid painful procedures such as branding and ear tagging. Allow the animals to explore without pushing them through too quickly. Walking young animals quietly without catching them is recommended as their first experience in the chutes. Quietly acclimating new heifers to walking through the chutes with no procedures being done may improve conception rates.

Nonaversive procedures such as weighing and sorting should also be avoided during the first experience in an effort to avoid the formation of fear memories associated with the chute. Place tasty feed where the animals can easily eat it after exiting the chute as an incentive for them to enter the chute the next time around.

For dairy cows and goats, acclimate young animals to head stanchions or milking stands before they start lactating. Feed them in the milking facility so they associate it with a positive experience. Avoid all painful procedures in the milking areas. Do not use milking stalls to restrain animals for veterinary procedures.

Moving Animals between Pastures

After the animals have moved through the gate, you should wait there until they have stopped to turn around and look at you.

Moving cattle on pasture can involve a continuum of techniques, from hardwired instinctual behaviors to completely trained animals led by a handler.

On ranches using intensive rotational grazing, you can teach mother cows, goats, and ewes to come when called, and a person or a vehicle can lead them to new pastures or the corrals. This completely voluntary movement is not stressful to the animals when done correctly.

Animals can be trained to respond to a specific call or horn, and not just to the sight of a vehicle or person. Nothing is worse than babies being left behind while the mothers hungrily chase your truck around a pasture while you are fixing fences. This is very stressful for the calves and lambs and can slow down weight gain.

If you blow the horn for a few seconds before putting out feed, the animals will learn to associate the horn, rather than the sight of the vehicle, with being fed. This keeps the cows from following the truck when you are trying to do other chores.

Always control livestock movement to a new location. *Never* allow animals to run into a new pasture or out of the old one. Drive or walk close to where they are grazing before calling them. This will help prevent calves becoming separated from their mothers. A vehicle or a person should either stay in front and lead the animals through the gate or park at the gate to control movement.

Movement can also be controlled by having the herd stop before the gate is opened. Do not open the gate until the herd is calm. If you open the gate to a new pasture when the herd is mobbing you, you are rewarding bad behavior. The animals need to learn that they have to "ask politely" and you will open the gate.

LEADER ANIMALS

Using calm leader animals is another excellent method to quietly move cattle or sheep. You can move an entire herd when the lead animal follows you with a feed bucket. Teaching animals to lead works especially well with small herds or flocks.

The cowboys who worked on the great cattle drives in the 19th century recognized the value of a good lead animal. An animal that would easily follow and that exhibited the leadership qualities of calmness and fearlessness was used year after year. Nervous, excitable cows that became leaders were culled.

You can recognize the natural leader of a group when cattle are moving by themselves to the water trough. A completely tame leader cow or steer that trusts you can provide excellent low-stress movement between pastures and into the corral. You can teach this leader animal to walk calmly to you when you call her. The other cattle will follow the leader at a walk, and this will decrease excited running, which can be highly stressful to young calves.

In his book *Grass-Fed Cattle* (Storey Publishing, 2006), Julius Ruechel suggests choosing your lead animal by letting him choose you. If you sit quietly, the cattle will come up to you. Gradually you will be able to start stroking the leader and he will become your friend.

Babysitter Moms

On large pastures, you'll observe that cows or goats will leave their babies with a female babysitter. She watches a group of small offspring while the other mothers graze. This babysitting behavior explains why a cow or doe goat is not always with her baby.

Never move the mothers when the babies are with the babysitter. This will cause stress because the babies will have a much harder time finding their own mother. Babies will usually be with their mothers at dawn and in the late afternoon, so those are safer times to move the herd.

Tips on working with the leader:

- To teach the leader good manners and not to push against you, take away the feed or stroking within one second after she pushes.

- Never hit the leader.

- Reward the leader with treats and stroking when she respects your space and does not push.

- Never open a gate for her if she is pushing against you. Wait until she backs off slightly, then open the gate. This rewards polite behavior.

Trained leader sheep or goats work extremely well for bringing sheep out of pastures and large pens. A good lead sheep can go into a large paddock and lead strange sheep she has never seen before out of the pen. Trained leaders can move sheep through stockyards, on and off trucks, and into meat plants. Unlike cattle, sheep have such intensive following behavior that they are more willing to follow a strange animal than other species would be.

Another good system is to mix in a few tame cows or steers with the younger cattle. This will help quiet them down and teach them where the water troughs are.

Tina Williams (Bud Williams's daughter) explains that animals should also know how to respond to being driven. Animals that have not learned how to react calmly to being driven may become agitated and stressed when they go to an auction or a packing plant. Before animals leave your farm, they need to learn how to calmly respond when they are being driven in and out of a pen, through a gate. They must learn how to walk calmly past a person and through a gate.

ORDERLY CATTLE MOVING IN THE SAME DIRECTION. A group of loosely bunched cattle being quietly driven in the manner of cattle handling specialist Burt Smith. When Burt moves cattle, he begins by working them into a calm, orderly loose bunch. When this is successful, all the animals are moving in the same direction. The orderly loose bunch must be formed before additional pressure is applied to the collective flight zone to move the group forward.

Training Tame Animals

Cattle, pigs, sheep, or goats in small operations often became very tame. They can be easily trained using operant conditioning methods. In her book *The Backyard Goat* (Storey Publishing, 2011), Sue Weaver has an excellent chapter on clicker training. This method was originally developed by animal behaviorist Karen Pryor.

In clicker training, the animal learns to associate a click sound (from a handheld clicker) with food. To teach the animal, the first step is to feed a treat *immediately* after a click. Do this about 10 times to solidify the association.

The advantage of the technique is that it enables precise timing of the reward stimulus. It can actually help prevent animals from mobbing a person at feeding time because the animals learn that they do not get the feed until they have heard the click.

Dangerous mobbing of people during feeding or changing pastures can also be prevented by teaching animals that the gate will not be opened or the feed will not be put out until they stand still. You want to reward good, polite behavior. If feed is put out or the gate is opened when animals are mobbing you, you have rewarded bad behavior.

Feeding treats can also be used to train animals to remain still during milking or veterinary procedures. Give the animal a treat when he stays still. If an animal is really unruly, give a treat if he stays still for even two seconds. In the beginning, give a reward to the animal if he does what you want for even an instant. Then gradually increase the time he has to stand still in order to get the treat.

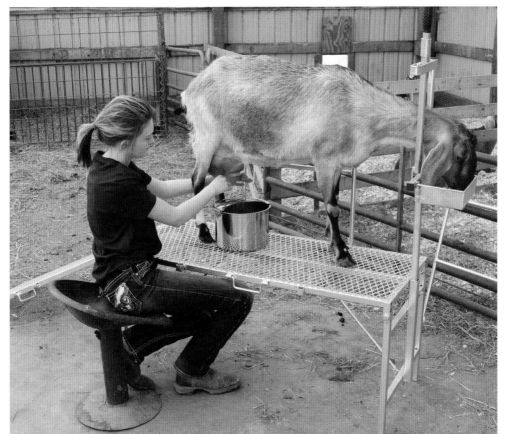

A milking goat on a milking stand. Your goat should be carefully habituated to the stand before she freshens and starts milking. Make a young goat's first experiences with the stand positive by feeding her treats.

How to Reduce Pushing Behavior

You can reduce pushing behavior in tame cattle by encouraging a submissive posture.

1. Stroke the animal under the chin and neck, and offer treats to get him to point his nose straight up toward the sky. This submissive stance is the posture cattle use when they are nursing, and it triggers instinctive passive behavior. It is a species-typical instinct.

2. When the animal puts his nose in the air, don't pat it — stroke it. If he bunts or pushes at you, instantly withdraw the treat until he understands he will get the reward only as long as he is being passive.

Cattle can be stroked under the chin or on the neck, withers, or back. They also love a good butt scratch. Have them take the submissive treat-taking posture periodically, to help them mind their manners when you lead them from pasture to pasture and to minimize mobbing behavior.

A correctly tied, easily adjustable rope halter

To reduce pushy or aggressive behavior, feed a treat with your hand held high and in such a way that the animal has to point her head toward the sky. This submissive stance is the posture calves use when they are nursing, and it triggers instinctive passive behavior.

TEACHING ANIMALS TO LEAD WITH A HALTER

If you plan to show an animal, you must train him weeks beforehand. The worst thing to do is to attempt to train an animal to a halter shortly before the show. Young animals will be easier to train than older animals. Animals that already tolerate being touched and having their body handled will also be easier.

Since the halter is something new, allow the animal to see it and sniff it. Adjust the halter as shown in the illustration above.

When using a rope halter, take care to avoid putting it too low on the nose. If it is too low, it can make it hard for the animal to breathe if he pulls back hard. One way to accustom an animal to wearing a halter and accepting pressure on a lead rope is to attach a short lead rope and let it drag. When he steps on the lead rope, he will feel pressure.

The big mistake that people make when training an animal to lead is to keep

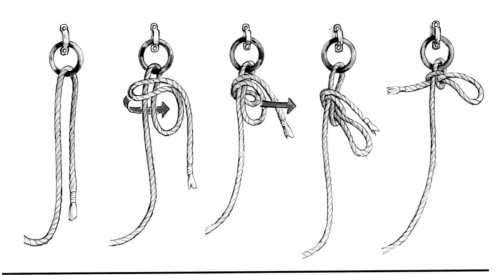

QUICK-RELEASE KNOT. 1. Run the rope through a tie ring or over a rail. **2.** Fold the free end of the rope and pass it around and behind the rest of the rope. **3.** Bring it back through the loop formed and pull till the knot is snug. **4.** Slide the knot up to the ring. **5.** To release, pull the free end.

pulling when he moves forward. When he steps forward, *immediately* release pressure on the lead rope to reward him for cooperating. Even if he just leans forward and starts to take a step, reward him with release of pressure on the lead rope.

To lead an animal with a halter, you also need to know how to tie a quick-release knot. This should be used when an animal is tied to a fence or post. In an emergency, a quick yank will release the rope.

Rotational Grazing without Fences

Progressive ranchers on extensive western ranges have learned to manage their cattle grazing without using fences. Shepherds in traditional rural societies who live on the range with sheep and goats have always done this. They constantly move

their flocks to new pasture. Rotational grazing is possible using a combination of principles defined in the previous sections and a good knowledge of your land.

To get the program started, you must move the herd a *considerable* distance from the original site. It is very important to move the herd far enough away so they won't wander back. Move the herd quietly at the speed of the slowest animal. Chasing animals out of an area does not work because frightened cattle are motivated to return to the previously safe place.

Tina Williams explains that cattle are more likely to stay in a new place if the handlers remain with them until they start grazing in random orientations. This gets their brains switched from the mode of being moved by people to the mode of seeking food and eating.

Young animals you've raised yourself are easier to train to graze in a chosen location than are a group of mature animals from several ranches. Old cows may have learned bad habits that are difficult to change. "Bunch quitters" that run off by themselves and hot-tempered cows that disturb the entire herd should be culled.

Both research and practical experience have shown that low-stress herding methods are effective for placing cattle on hilly highland pastures that need to be grazed and keeping them off fragile riparian lowlands near a stream. Both Derek Bailey, formerly at Montana State University and now at New Mexico State University, and Floyd Reed, from the U.S. Forest Service in Delta, Colorado, have done extensive research on herding methods to control where cattle graze. They have learned ways to make the Bud Williams methods of low-stress herding even more successful. These methods of cattle grazing also are being used successfully in the rough, hilly country of Colorado and Montana. Herders riding through the cattle every day teach them to be calm and easy to handle.

Calming Wild Cattle Entering Pens

No matter what type of grazing animals you are raising, you will need to know how to move them in and out of pens for bedding down for the night or for giving birth. You also will need to sort them for veterinary or other procedures. Cattle raised under conditions that dictate little or no human contact can be taught to tolerate people when they are brought into pens or small pastures.

Nebraska veterinarians Lynn Locatelli and Tom Noffsinger developed a method to reduce the size of the collective flight zone by gently penetrating the flight zone and then backing away at the first sign of a reaction. If there is no movement when reentering the flight zone at the same distance, back off. The principle is to reward the animal for not moving. The next time, you can get closer. Using this method, cattle gradually allow people to get closer to them.

Lowlands or Highlands?

Cattle have a preference for lowlands or highlands, influenced by both genetics and learning. Breeds of cattle originally developed for use in the mountainous terrain tend to prefer the hills. Derek Bailey found that Salers and Tarentaise cattle prefer the hills and Herefords prefer the lowlands. Ranchers have observed cattle that prefer hilly terrain may be more flighty and protective at calving compared with lowland cattle.

There is also a big difference in grazing preferences among individual animals within a breed. By tracking individual cows with a global positioning system (GPS), researchers found that some animals are definitely hill dwellers and others prefer the lowlands. Cattle that prefer the lowlands are more likely to graze over fragile riparian areas along rivers and streams.

Another method that Locatelli and Noffsinger use entails walking with groups of pacing, anxious cattle in the same direction as cattle movement to slow down the animals. This method is especially useful for calming a pen of freshly weaned, bawling calves. Bud Williams also has successfully employed these methods to reduce the stress of calves arriving at a feedlot.

Wild cattle are more likely to mill and pace in a large pen. Some stock handlers have found that putting the animals in a long, narrow pen stopped the pacing and the animals drank from the water trough more quickly than when confined in a large square pen.

Livestock handlers who routinely make careful observations of their herds and practice low-stress management techniques will be rewarded with calm, cooperative animals — allowing for safer, more productive livestock operations.

TIPS FOR GRAZING SUCCESS

• Move cows and calves to new grazing locations later in the day — around the same time calves bed down for the night. A herd will stay where the calves bed down. On pastures with fences, it may be preferable to move the cows in the morning, depending on pasture conditions.

• Do not let cows and calves get separated during movement to a new pasture. Move cattle at the pace of the slowest animals.

• Both sheep and cattle prefer the forage on which they were raised. Cattle and sheep raised on lush lowland pastures may be more difficult to move off these areas than animals born and raised in the hills. Cull problem animals that insist on grazing in the wrong place.

• Supplements are useful for improving the effectiveness of herding. Tasty molasses supplements should be placed at the new grazing site. Cows must have previous experience with the supplements and be attracted to them. Desirable supplements help motivate the cows to stay on the new site. Supplements should be put on the roughest hilly range to encourage them to go up there. To move cows along a ridge, gradually move the supplements.

• An unexpected research finding was that older, more experienced cows are more willing to graze on rougher terrain as compared with young heifers.

• During the hot summer weather, more water sources are needed in the hilly country to keep cattle out of the lowland riparian areas.

4 | MOVING ANIMALS THROUGH CHUTES AND HANDLING FACILITIES

These different views show handlers standing in the proper position to move animals into the single-file chute. Sheep naturally go back to where they came from. In the layout shown at lower left they circle around the handler.

By understanding both the instinctive and learned behaviors of grazing animals, a producer can help animals experience procedures such as vaccinations and pregnancy checking with very little stress. Appropriate handling methods and practices that consider both animal and human welfare are central to a low-stress regimen.

This chapter contains proven techniques for handling that use the principles of flight zone and point of balance to ease movement of animals through systems of single-file alleys and squeeze chutes. Simple fixes such as removing visual distractions make facilities less frightening, more efficient, and safer. By controlling what an animal sees, hears, and experiences, producers can protect their investment and alleviate the psychic wear and tear on the animals.

PRINCIPLES

Animals naturally follow a leader.

Move cattle, pigs, and goats in small, separate bunches through crowd pens and single-file chutes.

Sheep can be moved in a continuous flow, similar to siphoning water, because they have intensive following behavior.

Animals will move toward a lighted area; however, they will not approach blinding light such as a rising or setting sun.

Remove distractions from corrals and chutes that cause animals to balk.

Animals have a natural inclination to go back to where they came from.

Be careful with the lone animal that has become separated from its herdmates. A frantic, lone animal is a major cause of injuries to both animals and people.

Livestock Handling Principles Inside Corrals and Chutes

There are a few basic principles that livestock producers and handlers on farms will want to employ for ease of movement. Problems can typically be traced to a lack of handler training or a design problem. Even though a few extra pigs jammed in the crowd chute or a couple of jolts of a cattle prod may seem insignificant to some people, they can make animals severely stressed. You must get inside the mind of the animals to understand how neglected details can both slow down handling and cause fear. The goal is to move animals at a walk or a trot and prevent them from running.

MOVE SMALL BUNCHES OF ANIMALS IN CROWD PENS AND SINGLE-FILE CHUTES

Cattle, bison, pigs, and goats should be moved through handling facilities in small, separate bunches. Don't try moving too many at a time. Good handling requires more walking. When market-weight pigs are loaded onto vehicles through a narrow alley that is less than 36 inches (91 m) wide for transport to a processing plant, they should be moved in groups of three to five. Small weaned piglets and nursery pigs can be moved in groups of up to twenty. For sows, the maximum group size to be moved at one time is five. On small farms, moving one sow at a time is often recommended. A single sow will respond easily to the handler positions shown on the flight zone diagram in chapter 3.

When cattle or bison are being handled, the crowd pen leading to the single-file chute should never be more than half full. If the crowd pen holds ten large animals when it is full, put in only five animals at a time. Take care never to push up the crowd gate tightly against the animals, or they will not be able to turn to go into the chute.

Sheep are the one exception to the small-group rule. Whereas most grazing animals should be handled in small bunches, sheep can be moved in large groups to maintain continuous following behavior. Moving sheep from a crowd pen into a chute is like siphoning water. You never want to break the flow of the follow-the-leader instinct.

USE THE FLIGHT ZONE TO MOVE ANIMALS THROUGH SINGLE-FILE CHUTES

As with taking advantage of instinctive behaviors to herd livestock in and out of pastures and pens (discussed in chapter 3), handlers calmly using similar patterns can easily move animals through chutes. For example, to move animals *forward* in a single-file chute, the handler moves at a fast walk inside the flight zone while going in the *opposite* direction of desired movement. When the handler crosses the point of balance at the shoulder, the animal will move forward. This simple movement is much less stressful than using an electric prod or slapping to induce animals to move forward. To slow down forward movement, the handler should stay outside the flight zone and walk in the same direction as the desired movement.

For cattle, goats, pigs, or bison, fill the crowd pen only half full. Good handling will require more walking to bring up small groups of animals. When market-weight pigs are being loaded onto a trailer, move only three to five at a time. When small pigs, up to 20 pounds (9 kg), are moved, larger groups, up to twenty, can be used. Note in the figure that the crowd gate is not pushed up tight against the cattle. A common mistake is attempting to push animals with the crowd gate.

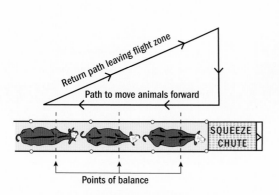

A crowd pen can be filled completely with sheep, due to their intense flocking behavior; their intense following behavior causes sheep to work in a continuous flow.

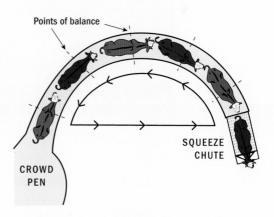

HANDLER MOVEMENT PATTERN TO INDUCE CATTLE INTO A SQUEEZE CHUTE

Cattle will move forward when the handler, walking inside the flight zone, passes the point of balance at the shoulder of each animal. The handler walks in the opposite direction of the flow of the animals alongside the single-file chute. A larger version of this pattern can be used in pastures and in large pens.

HANDLER MOVEMENT PATTERN TO MOVE CATTLE THROUGH A CURVED CHUTE

Cattle will move forward when the handler crosses the point of balance of each animal.

TAKE ADVANTAGE OF NATURAL FOLLOWING BEHAVIOR

To maximize the flow of cattle, goats, or pigs entering the single-file chute, wait until the chute is nearly empty before refilling it. Because there is sufficient room in the chute for four or five animals to enter when it is nearly empty, a handler refilling the chute at that point takes advantage of the animals' natural following behavior. Attempting to force one or two animals at

a time into an almost full single-file chute often causes balking and turning back. The crowd pen works more efficiently if used as a "passing-through" pen; cattle and pigs enter the single-file chute to a loading ramp or squeeze chute with the greatest ease if they are kept moving through the crowd pen without stopping. When they stop in the crowd pen, they often turn around in the wrong direction.

A live decoy sheep will attract flocks of sheep into a chute. Place the decoy ahead in a small pen located just outside the exit of the single-file chute where approaching sheep can see it. To keep the decoy from becoming stressed, do not use the same sheep all day to guide the others. Another alternative is to use three decoy sheep so that the decoys have company.

Cattle and sheep facilities should be designed with relatively long single-file chutes to take advantage of following behavior. It is important not to get the chute too long, but it should have sufficient space so that four to six animals will fit in it. A single-file chute that holds only one or two animals works poorly because it is impossible to efficiently use following behavior. A double-file chute (see the illustration opposite) is especially useful if space

Avoid Too Much Pressure on the Flight Zone

When handlers are working livestock with a large flight zone in an enclosed space, such as a drive alley, chute, or crowd pen, they must take great care to avoid penetrating the flight zone too deeply. This can result in extreme fear, jumped fences, and animals turning back on a handler. If animals in an alley start to turn back, you must back up and get out of the flight zone. Backing out of the flight zone will often prevent the animals from turning back and running by you. If an animal rears up in a chute when you deeply penetrate the flight zone, you should back up and remove yourself from the flight zone. Attempting to push the animal down may increase both fear and agitation.

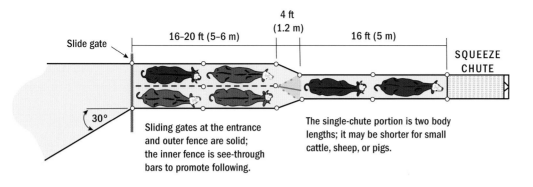

A double-file chute system is the best design choice if a straight chute is needed: for example, if space is limited. The outer fences of the alley are solid, but the inner fences are solid on the bottom half but see-through in the middle. The animals can see other animals next to them, which works well for sheep, cattle, and pigs.

is limited. A double-file chute holds cattle in two single-file side-by-side alleys. This system also encourages following behavior. When one animal moves forward, a herd mate standing beside him usually follows.

Cattle, pigs, and sheep will wait quietly in a single-file chute. Bison become severely agitated while standing in line, however, and may gore each other. Bison facilities work best if the single-file chute is shorter and contains only one or two animals at a time. Keeping bison separated in the chute with a series of sliding gates that form single animal compartments has limited success for producers. The gates prevent injuries. However, by the time the fifth or sixth bison reaches the squeeze chute, he will be highly fearful and agitated.

USE THE PROPER AIDS FOR DRIVING ANIMALS

A flag or paddle quietly moved along the side of an animal's head can be used to turn the animal. The handling aid is used to quietly guide the animal and should not be used to hit it. A light stick 6 to 8 feet (2 to 2.5 m) long, with a plastic grocery bag on the end, is a useful and inexpensive tool for guiding cattle or bison in crowd pens. Commercially available flags on light flexible shafts are also good tools. Handlers must avoid waving the stick and becoming aggressive with it. Just a bit of pressure in an animal's flight zone works very nicely and usually keeps the animal from becoming stressed or scared the way it might when a cattle prod is used.

Solid plastic panels work well for pigs. A pig usually will not attempt to run through a visual barrier. Panels are heavy, however, and carrying them can be tiring. A "witch's cape," a piece of lightweight plasticized cloth that is stiffened along one edge, is a good pig-driving tool and easier to hold for long periods. Most of these tools can be made at home or purchased at a feed store or farmer's exchange.

Some people highly skilled at low-stress handling of animals prefer not to use a driving aid. They have learned how to use very small movements of their bodies to get animals to respond. The big mistake people make is yelling, whistling, and arm waving. **When driving animals, try to concentrate on using the smallest possible movements of your body to induce the animals to move.**

Livestock Driving Aids

Use the electric prod only when an animal absolutely refuses to move. Never use an electric prod as a primary driving aid. Use it only if an animal absolutely refuses to enter the squeeze chute. In the case of a complete refusal, an electric prod is preferable to hitting, tail twisting, or poking a sensitive part of the animal. Acclimating young animals to moving through the chutes with no procedures being done will help prevent refusals. Do not use electric prods on infant animals.

The worst place to keep an electric prod is in someone's hand. A prod should be placed where it is easy to get to, but pick it up only when absolutely needed. After getting a stubborn animal to move, put away the electric prod. Observations at many facilities indicate that handlers' attitudes about the animals improve when they stop carrying electric prods.

In an ideal world, the electric prod would be banned from ordinary use. But there are a few situations in which the electric prod is the very best tool. You might be getting gas at a truck stop, for example, and discover that a steer has fallen down in the trailer. You want to get the animal up for its own safety and the safety of the other animals, but poking at it to get up might be unsuccessful. You can't open the trailer and risk having cattle or other animals racing across the highway. At that point, pressing the electric button is absolutely necessary and generally effective.

An electric prod delivers the same shock as a live electric fence. Poke the animal only on the rear end if you have to use it. No device should ever be applied to the rectum, vagina, eyes, ears, nose, udder, or other sensitive parts of an animal. Electric prods should not be used on sheep. Wool is an insulator and the parts of the sheep that are not covered by wool are sensitive areas that should never be poked.

Farm managers should readily fire anyone who abuses an animal. My observation is that some people have an innate talent for working with animals. Unfortunately, there are others who become too impatient and do not work well with animals. If handlers really work thoughtfully with grazing animals, they won't have to use the prod at all. If producers are using the prod more than once or twice per 100 head of animals, they should seriously rethink their handling practices.

A handler must not carry around an electric prod. Better to use a flag tool at all times instead. The prod should be picked up only if an animal absolutely refuses to go into a squeeze chute or stun box. A momentary shock is preferable to physical abuse.

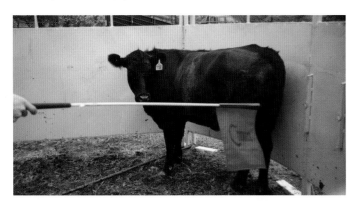

Flag for moving livestock in the crowd pen. When the flag is moved slowly, the animal will turn and look at it.

A rigid board or panel is effective for moving pigs. Pigs respect a solid barrier, although after some time it can become a heavy burden for the animal handler. A solid panel must be used to move mature boars. Handlers need the added protection if a boar decides to attack.

A "witch's cape" made of plasticized cloth cut to the width of the alley and attached to a stick across the top can be used to move sows, piglets, and sheep out of a big pen or down a wide alley. In narrow alleys, use half a witch's cape.

A large flag on a short stick works well for moving pigs and other small animals. The flag is made from a lightweight plasticized cloth. In many situations, the lightweight flag can replace a rigid panel. The flag is moved quietly behind the animals and should not be waved vigorously.

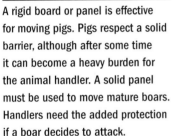

Some people prefer to use a fiberglass stick with a plastic paddle attached to move larger animals. It should not be used to hit an animal. Cattle in a chute will often move forward when their back is firmly stroked by hand, from their neck toward their rear.

USING A FLAG TO TURN AN ANIMAL. A plastic bag or plastic tape attached to a stick works well as a flag for moving and guiding animals. Slight movement of the flag by the animal's head will turn him. Never wave or shake the flag. When animals are calm, they will respond to very slight movements.

All of these sheep have their ears pinned back, indicating that the dog is being too aggressive toward them.

A well-trained dog calmly holds a flock of sheep. The dog has retreated back into the pressure zone and removed himself from the flight zone. Most of the sheep are turning and facing the dog. This is a sign that he has backed out of the flight zone.

USING DOGS TO WORK ANIMALS

Well-trained Border Collies and Kelpies can do an excellent job of low-stress gathering of sheep on a pasture. Never allow dogs to bite sheep, however. Research by New Zealand animal behaviorist Ron Kilgour proves that being bitten by a dog is extremely stressful for sheep.

If dogs nip at cattle confined in a chute where they can't move away, the cattle can become dangerous kickers and injure people. This is especially perilous for people who may need to handle your cow or steer at an auction or slaughter plant.

To minimize stress, dogs must be kept away from squeeze chutes and other restraint devices. Many ranchers have banned dogs in their cattle corrals altogether. Dogs work best for herding grazing animals in open pastures or large pens, where the animals have room to move away.

Livestock Hide Their Pain

Prey species have a natural instinct to mask pain in order to survive predators. This is why a calf that hurts himself or has been subjected to a painful procedure may continue grazing. That livestock do not experience pain is a common misperception among handlers and producers. A severely injured calf or sheep may lie moaning when nobody is around but instantly jump up and appear normal when he sees a person.

MORE HANDLING TIPS

Small changes can make a big difference when working to achieve calm, stress-free animal movement.

Be careful with the lone animal. Grazing animals often become highly agitated and fearful when separated from their herd mates. A lone animal is more likely to injure himself or his handlers. Never get in a small confined area with an agitated lone cow or bison. A lone animal will calm down when herd mates are brought in.

Ease squeeze chute entry. Use the movement patterns described on pages 100 and 102 to induce cattle to enter the squeeze chute. This is often the most effective way to eliminate or greatly reduce electric prod use. Another method is to stroke the animal along the backbone from head to tail.

Give animals time to calm. Calm animals are easier to handle than fearful, excited animals. If the animals become agitated, 20 to 30 minutes is required for them to calm down. Take a break to allow scared animals time to settle. Frightened animals bunch up and are more difficult to separate or sort.

Change handler position by the crowd pen. If the animals refuse to enter the single-file chute, try positioning yourself on the other side of the crowd pen. This affects the direction animals move as they approach the chute entrance. On round tubs, a handler working at the crowd gate pivot with a flag can easily make cattle circle around him or her and enter the single-file chute.

Cattle will often enter the single- or double-file chute more easily when the handler is positioned near the entrance. In this position, the handler can make slight movements back and forth across the point of balance to either speed up or slow down the cattle as they enter the chute. To prevent balking, the person should move into this position after the animals have started to enter the chute.

Never Lift or Grab Sheep by the Wool

Grabbing sheep by their wool can cause severe bruises and pain. Lifting a sheep by its wool would be similar to lifting a person by his or her hair. If sheep are grabbed by the wool during transport or handling at a processing plant, large pieces of bruised meat may need to be trimmed. Bruised meat cannot be used for human consumption.

Facility Fixes That Alleviate Animal Stress

You'll be amazed at how a few little changes in the facility can drastically improve the movement of the animals. A distracting detail as tiny as a contrasting-colored floor mat in front of the chute or a brilliant patch of sunshine in the main alley can cause animals to balk or turn back. Not only must handlers allow herds to move slowly through the facility as a good handling practice, but they also must allow the lead animal time to investigate these odd-looking things if they want to prevent them from balking and turning back.

To improve animal movement, it makes good sense to inspect the entire facility and see what changes you can make to minimize distractions and fear. Common distractions that should be removed in order to improve animal movement are coats hung on fences, hoses on the floor, and vehicles parked by handling facilities.

BLOCK VISION TO KEEP ANIMALS CALM

Several studies show that blindfolding with a completely opaque cloth can have a calming effect on cattle, because cattle do not fear what they do not see. For them and other grazing animals, it's out of sight, out of mind. The movement of wild, large-flight-zone cattle, bison, deer, or sheep through a handling facility can be improved when solid sides are added to

YOU MANAGE WHAT YOU MEASURE

A good livestock producer measures performance indicators such as weight gain, calving ease, and calving percentage. It is also recommended to measure your performance in handling. This will help ensure that you do not get into bad habits without realizing it. Too often a person will attend a low-stress handling workshop and become really enthusiastic. Over a period of a few months, however, that person may slowly revert to harsher ways of handling and be surprised when reviewing measurements.

The following are some simple variables you can score to keep animal handling low-stress:

- Percentage of animals falling during handling (body touches the ground)

- Percentage of animals stumbling

- Percentage of cattle, goats, or pigs that moo, bellow, or squeal during handling in a squeeze chute or restrainer before a procedure such as ear tagging is begun (not an effective measure with sheep)

- Percentage of animals that strike a fence or gate

- Percentage that run (rather than walk or trot) during handling

- Percentage moved with an electric prod (ideally none)

- Percentage caught in the wrong position ("miscaught") in a restraint device

The numbers will help you tell if you are getting better or worse.

existing open-sided fences. Solid sides prevent the animals from seeing people and other distractions outside the chute. It is especially important that approaching animals do not see people up ahead around the squeeze chute; sometimes installing a shield so that the animals cannot see the people is helpful.

I came to fully appreciate the tremendous influence that blocking vision has on livestock behavior when I worked on the design of a restraint at a meat plant. To the amazement of all involved, simply taping a large piece of corrugated cardboard over the top of a squeeze chute kept 1,200-pound adult bison from rearing. People who handle bison and deer have used solid sides on squeeze chutes for many years, yet this design is still not a convention in many handling facilities. Blocking vision has the greatest effect on wild cattle or bison with little or no prior contact with people. It is less effective on animals habituated to people.

Blocking vision to prevent animals from seeing people deep in their flight zone is especially important for animals that are wild and have a large flight zone. If animals have become completely tame and have no flight zone, blocking vision will have little effect. The use of solid sides on chutes is especially recommended if cattle are wild and not accustomed to being handled, or if they are being handled by unskilled or inexperienced people.

Solid sides on single-file chutes, crowd pens, or loading ramps are especially recommended when handling really wild animals or when handling is done by unskilled or inexperienced people who are not trained to stay out of the animal's flight zone. Open sides will work for gentle, tame animals or with highly skilled handlers. The most important fences to make solid are those on the *outer perimeter* to block the animal's view of distractions such as moving vehicles or people walking by.

Rubber louvers made from the strips of a conveyor belt facilitate cattle entry into the squeeze chute because they prevent an incoming animal from seeing people standing around the chute. Old round-baler belts work really well. The louvers are installed on a 45-degree angle.

A cow's-eye view into the squeeze attracts the animal toward a lighted opening. The rubber louvers on the sides prevent the approaching animal from seeing human handlers and other distractions outside the chute.

The Cardboard Test

Blocking vision to facilitate calmness is one way to use natural behavior rather than force to control an animal. Vision barriers are calming because they block distractions, prevent the animals from seeing an escape route, and impart a sense of security.

If you are not convinced that the result will be worth the investment, experiment. Cover the sides of the squeeze chute with cardboard before installing permanent louvers. The most important part to cover is the back half nearer the tail gate.

Taping cardboard over the top of the chute as well as the sides is recommended to keep bison from rearing. Bison squeeze chutes should also have a solid gate located 3 to 4 feet (1 to 1.2 m) in front of the head gate. The material does not need to be sturdy. It just has to effectively block vision. Do not use thin plastic or other materials that might flap and frighten the animals.

A dark box chute is very effective for artificial insemination and pregnancy checking. The cow stands in a box with solid sides, front, and top. There is no head catcher or squeeze chute. She stands still because the entire box becomes a blindfold.

ON RESTRAINT DEVICES, CREATE SOLID SIDES AND LIGHTED WAYS

A kosher slaughter restrainer I designed used the principles of blocking vision and light to attract an animal to walk in quietly without attempt to lunge at the head opening. The design of this box calls for completely solid sides to block peripheral vision and prevent the animal from seeing the rabbi and other activity outside the restrainer. The head opening is lit up with diffuse indirect lighting to encourage the animal to voluntarily put his head through. Animals can see only the wall and a lighted space in front of the box. The cattle at the meat plant where this was installed were much calmer than cattle entering a conventional squeeze chute with open bar sides. The operator of the manual squeeze has to stand inside the flight zone to operate the chute. A piece of cardboard installed on the back half of the open bar sides will facilitate cattle entry. It prevents incoming cattle from seeing the chute operator.

Any material, provided it is not loose and flapping, can be used to create vision barriers. Sheet metal, plywood, planks, and old conveyor-belt material are all commonly used. Louvers made from rubber conveyor belting 6 to 8 inches (15 to 20 cm) wide can be installed on the drop-down bars on a 45-degree angle. The bars can still be opened, but incoming animals cannot see out as they enter.

ALLOW ANIMALS TO APPROACH LIGHT

All species are sensitive to the illumination in a handling facility and have a strong tendency to move from dimly lit areas to brighter areas as long as the light is not shining directly into their eyes. On sunny days, an animal will often refuse to enter a dark building. The easiest way to encourage animals to willingly enter is to lay out your handling facilities so that approaching animals can see daylight through the opposite side of the building. It works best if they can see through the building.

The wrong lighting can ruin the performance of a well-designed facility. New buildings constructed over cattle and sheep handling facilities should be equipped with white plastic side panels or skylights to let in light. White translucent panels work well to let in plenty of natural light and to eliminate shadows. At night, lamps can be used to attract animals into buildings, trucks, or chutes. Even with electric lighting, animals may be reluctant to enter buildings that are darker than the outdoor daylight. On bright sunny days, the lamps may have little effect because sunlight is brighter than artificial lighting.

Animals will not approach blinding light. Squeeze chutes and loading ramps should be oriented so that they do not face into the sun during handling times. In the early morning, a loading ramp oriented in an easterly direction would put the early sun in an animal's eyes. The same orientation, however, would have no effect if animals were being loaded at midday or late afternoon. As the sun moves through the sky, changing light may influence handling. When building a facility, it is a good idea to think about the time of day that animals are loaded onto trucks or worked through chutes.

Unlike their sheep and cattle cousins, deer and elk prefer a dimly illuminated building with no skylight. In Australia and New Zealand, deer are handled in a dark building with one very dim light. Deer that have a big flight zone outside will often allow people to walk up to them in the dark building. In this situation, the entire building becomes a blindfold. Over the years, deer have been bred to be more docile. A dark room may not be required with tamer genetic lines of deer.

REMOVE DISTRACTIONS

When animals balk and refuse to enter a chute or walk down an alley, it's often caused by small distractions that people fail to notice. Shadows, reflections on water, and detecting movement up ahead are all examples of the small sensory details cattle react to in their environment.

An experienced dairy cow will ignore a drain she walks over every day, but the same drain may cause a young heifer or doe goat that has never seen it before to balk or turn back. Calm animals show you the things that distract them by stopping, looking at, and pointing their nose and ears at the object or streak of light. If there is a distraction in a facility, such as a drain on the floor, the lead animal may stop and put its head down to investigate it. The handler must wait for the leader to raise its head back up before attempting to move the animals forward.

Before working animals through a handling facility, it is a good idea to walk through it and remove or modify the distractions that cause balking. This will make it easier for you to move the animals through. Sometimes it is helpful to bend down and put your eyes at the animal's level to be able to see what he may be seeing. For sheep and goat facilities, you will need to get down on your hands and knees.

HEAD POSITION Ron Gill at Texas A&M University explains this simple principle of head position.

HEAD DOWN with its nose on the ground: the animal may balk and refuse to move.

HEAD NORMAL: the animal will move easily.

HEAD RAISED HIGH: the animal is scared and will not move easily.

Grazing animals may balk at a bright spot of sunlight or dark shadows in a handling facility. Both the animal's ears and nose are pointed at this sunbeam. Calm animals will show you the location of a distraction. (Note that the animal in this illustration is stepping out onto a mat made from either woven tire tread or woven old round baler belts — excellent nonslip flooring.)

It is good to crouch down and get an animal's-eye view in a chute with open sides. Visible people, vehicles, movements, or shadows may cause balking. Solid sides, especially on the outer perimeter of a facility, will reduce these distractions.

Remove dangling chains; they will cause animals to balk. Calm animals will often look directly at a distraction such as a chain and show you where it is.

Design of Livestock Ramps

These three drawings illustrate correct and incorrect cleat spacing for loading ramps, or a ramp onto a milking stand for a goat. The correct cleat spacing enables the animal's hoof to fit comfortably between the cleats. If the cleats are spaced too far apart, the animal will slip. If they are spaced too close together, the animal's hooves may slide over the top of them. The correct spacing is 8 inches (20 cm) of clear space between the cleats for cattle, horses, bison, and other large animals. Smaller animals such as sheep or goats will require closely spaced cleats with 3 to 4 inches (8 to 10 cm) of clear space, depending on hoof size. For all animals, metal rods 1 inch (2.5 cm) in diameter work well.

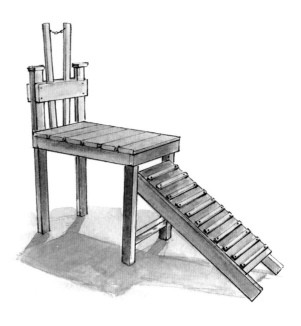

LOADING RAMPS should be well maintained. This ramp has broken cleats, and these piglets will have difficulty climbing this ramp. Broken cleats must be replaced.

CORRECT CLEAT SPACING: the hoof fits easily between the cleats. If the cleats are too far apart, the hoof may slip. If the cleats are too close together, the hoof may slip over the top of them.

TROUBLESHOOTING DISTRACTIONS

Distraction that causes animal to balk	Solution to improve animal movement
Sparkling reflections on puddles, smooth metal, or vehicles with reflective paint	In indoor facilities, move overhead lamps to eliminate reflections. Outdoors, use nonreflective paint to cover metal sides or parts. Work livestock at a different time of day.
Chains that jiggle	Remove them.
Metal clanging or banging	Install rubber pads.
High-pitched noise	Choose equipment that emits less high-pitched noise.
Air hissing	Silence with mufflers or pipe outside.
Drafts blowing at approaching animals	Change ventilation fans.
Clothing hung on fence	Remove it.
Piece of moving plastic	Remove it.
Fan-blade movement (when fan is turned off)	Turn on the fan because when the fan is turning fast, the animals cannot see the movement. Another option is to turn the fan so it no longer faces approaching animals.
Seeing people moving up ahead	Install a solid shield.
Small objects on the floor, such as coffee cups	Remove them.
Changes in flooring and texturing	Cover the floor with sand, dirt, or straw.
Drain grate on the floor	Install drains outside the area where cattle are worked.
Sudden changes in the color of equipment (colors with high contrast are the worst)	Paint all parts the same color.
One-way backstop gate	Move backstop two or three body lengths away from the entrance of the single-file chute. Another good method is equipping the backstop with a remote-control rope so that it can be held open until the chute is filled. Another option is replacing a backstop located at the single-file chute entrance with a sliding gate.
Visible vehicles that are moving or have high-contrast colors such as yellow or bright blue	Remove them or install a solid side.
Metal pipes on the ground that hold the facility together	Cover the pipes with dirt so they match the surrounding ground.

Restraint Practices

Cattle with large flight zones can become severely stressed and agitated in a squeeze chute with open side bars. This distress is often due to a deep invasion of the animal's flight zone by the operator or other people standing close by who can be seen through the open barred sides. People who are giving vaccinations should stand back outside the flight zone. After the animal is restrained, they should step up to the animal to do procedures and then back away.

All restraint devices and scales must have nonslip flooring. All species of animals may become agitated if they are standing on a slick floor where they make many small repeated slips. Animals are often reluctant to enter a scale or restrainer if the floor moves and jiggles. Allow the animal time to investigate the movement.

Animals should walk or trot into and out of a squeeze chute or sheep restrainer. When cattle or bison run fast into a squeeze chute, they can sustain shoulder and neck injuries if they hit the head gate too hard. Quiet handling pays in dividends by reducing injuries and weight losses caused by shoulder and neck pain. One large Colorado feedlot reported reduced sickness when it began handling animals more gently in the squeeze chute.

When animals enter the squeeze chute, pressure should be applied slowly instead of suddenly, in an attempt to "catch" the running animal. Slow, steady movement calms the restrained animal.

Sudden, jerky motion of the apparatus causes agitation and discomfort. When an animal's vision is blocked, he will stand and allow his head and body to be positioned in the device.

Padded restrainers are often used for deer and elk. Poorly designed restraint devices, where the floor is suddenly dropped, are frightening to the animals. A drop floor should be used only if padded sides securely hold the animal when the floor is lowered.

Cattle seldom resist pressure from the apparatus when it's applied slowly and with optimum pressure. **Optimal pressure** is defined as the ideal amount of restraint pressure on each individual animal. For that to happen, sufficient pressure must be applied to make the animal feel restrained while avoiding excessive pressure that will cause pain and discomfort.

If too much pressure is applied, the animal may vocalize his discomfort or strain to breathe. To resolve the problem, slowly lessen the pressure. A sudden release of pressure by the operator will excite the animal. Many people make the mistake of applying more pressure when an animal struggles. An animal will often stop struggling if the pressure is slowly, slightly reduced.

Conversely, if too little pressure is applied during restraint, animals are able to move around too much, making any procedure difficult to perform and making them highly agitated, thinking they can escape. The proper amount of pressure applied over a broad area of the body has a calming effect.

The best squeeze chutes have squeeze sides that close in evenly on both sides. This keeps the animal in a balanced position. Animals will often fight restraint if they slip or feel off balance. It's also important for the squeeze chute to be properly adjusted for different-sized animals. If the chute is not readjusted after handling small animals, larger livestock may have difficulty entering. Likewise, a squeeze chute adjusted for large livestock may not hold small animals securely.

Restrainers that invert sheep onto their backs are more stressful than are those that position the sheep upright. A restraint table that tilts sideways is preferable to fully inverted restraint. Sheep can be easily trained with tasty feed rewards to voluntarily enter a well-designed squeeze chute that tilts them on their sides. Cattle can also be trained to enter a comfortable tilting table for hoof trimming. If the belts or other parts of the restrainer pinch or hurt, however, it will be difficult to train animals to enter the device.

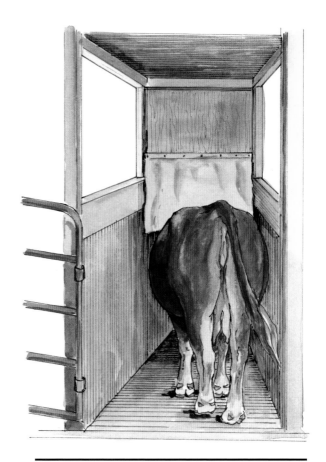

A dark box chute is simply a box made of wood or panels. It has solid sides, a solid door in front, and solid top that runs halfway to the middle of the animal's back. At the front is a small window.

Build a Dark Box for Artificial Insemination

Simple to build, a dark box chute works very well for artificial insemination and pregnancy testing. A dark box chute has no head gate or squeeze sides. It is a box with solid (often wood) sides, a solid front door, and a solid top that stretches half the length of the animal's back. To attract the animal into the dark box, make a small slit in the front at eye level, which provides the animal with a view that is not scary once she is in the box. The chute often features a canvas flap that is attached to the wood top and is flipped down onto an animal's back after she has entered. A simple chain secures the animal. After the procedure, the front is opened and the animal simply walks out.

PRINCIPLES OF LOW-STRESS ANIMAL RESTRAINT

Use the following principles to minimize fear and anxiety when animals must be restrained:

- Block vision with either solid sides or louvers to prevent animals from seeing people deep in their flight zone as they enter the squeeze chute. This is especially important for animals with large flight zones. It is not required for tame animals that allow people to walk up and touch them. Tame animals will voluntarily enter the head gate for a feed reward.

- Block vision of an escape route, but allow animals entering a restrainer apparatus to see a lighted area ahead. They will not walk into a dark space.

- Slow, steady pressure applied by a restraining device is calming. Sudden jerky motion causes excitement and agitation.

- A restraint device must supply sufficient pressure to provide the feeling of being held, but excessive pressure that causes pain must be avoided. If the animal is squeezed too tightly, reduce the pressure slowly. Sudden release causes agitation. Well-designed restrainers apply pressure to broad areas of the animal's body and can have a calming effect.

- Cattle will stand more quietly and remain calmer if they can see another animal within 3 feet (1 m) of them, but they may lunge and become excited if they see herd mates they want to join that are farther away.

- Install a solid roof on bison chutes.

- Use quiet handling in the chutes leading up to the squeeze chute. The animal should walk (not run) in and out of the squeeze chute.

- Use nonslip flooring. Animals get frightened if they start to slip, even slightly. Sometimes when an animal becomes agitated in a squeeze chute, it is due to a series of small, rapidly repeated slips. Adding a nonslip surface or a cleat to stop the slipping may help keep animals calm.

- A restraint device should support the animal in a balanced position. If the animal feels off-balance, he may struggle and become fearful.

- A restraint device must not pinch the animal.

- Never leave an animal alone in a restraint device.

Many problems with animals struggling in a restraint device are caused by animals becoming agitated and fearful in the lead-up chute. Calm animals are easier to restrain and will walk quietly into a restrainer. Often problems in restraint devices can be reduced by improving handling prior to restraint.

ADJUSTING SQUEEZE CHUTES AND RESTRAINT DEVICES

The use of a complete squeeze chute is strongly recommended for all cattle and bison that are not trained for head restraint. Restraint of the body will prevent an animal from fighting the head gate.

On hydraulic chutes, the pressure relief valve must be adjusted to prevent excessive squeeze pressure. Excessive pressure can cause severe injury such as a ruptured diaphragm or broken bones. If cattle strain or vocalize (moo or bellow) when the valve handle is held down until the hydraulic valve bypasses, the pressure setting is too high. Adjusting the pressure relief valve sets the maximum pressure that a hydraulic squeeze chute will exert. When the hydraulic pressure reaches the set pressure, the hydraulic fluid stops flowing to the cylinders that operate the squeeze and flows back into the reservoir.

The pressure relief valve should be set so that the chute will automatically stop squeezing before the bovine or bison strains, bellows, or has difficulty breathing. If a head gate has additional hydraulic-powered head-restraint devices such as neck extenders, these devices must have a separate pressure relief valve that is set to apply much lighter pressure than the rest of the squeeze chute. These devices must be used very carefully to prevent injuries.

Vocalizations to indicate pain are never evident when handling sheep. Detecting sheep stress at all is very difficult. Because sheep will not alert you when they are afraid or in pain, handlers should be particularly careful and observant. Pigs, however, are very expressive and will squeal at the slightest little distress. An occasional squeal can usually be ignored, but continuous loud complaint is a sign of distress that demands your attention.

A head gate with a curved-bar stanchion provides better control of the animal's head movement than does a straight-bar stanchion. To prevent choking in a head gate with curved stanchion bars, the head gate should be used with either a full-squeeze chute with the squeeze sides adjusted so that the V-shaped sides hold up the animal's body or a brisket bar to prevent the animal from lying down. The width adjustments should be changed when different-sized animals are handled. Pressure exerted by a curved-bar head gate on the carotid arteries, which pass up through the neck and supply blood flow to the head, can kill an animal quickly. If you ever see an animal in a curved-bar

When to Use a Straight-Bar Stanchion

The main advantage of a straight bar stanchion is an animal will not choke if it lies down. A straight-bar stanchion is the best choice for animal restraint in the following situations:

- If the head gate is mounted on the end of an alley with no full-body squeeze

- For chutes and head gates used to hold show animals for grooming

- To assist with calving

- To hold a cow during artificial insemination, pregnancy testing, or embryo transfer

- For sheep and goat milking stands

A straight-bar stanchion head gate is ideal for operations where an animal has to be restrained in the chute for procedures that take a long time. Body restraint is not required because in a straight-bar stanchion, the animal can lie down safely.

Curved-bar stanchions provide greater head control than straight-bar stanchions do, but they *must* be used with either a body restraint or a brisket bar to prevent the animal from lying down and choking.

stanchion collapse and start to shake, you must *immediately release* the head gate.

There are two basic ways squeeze sides work. They can either be hinged at the bottom to form a V shape that supports the body or remain vertical when the squeeze is tightened. Some veterinarians prefer a chute that does not pinch the feet together at the bottom because it allows cattle to walk in and out more easily. This is especially helpful for really large cattle.

A squeeze chute with straight squeeze sides must be equipped with either a straight-bar stanchion head gate to prevent choking or a brisket bar to prevent cattle from lying down. A brisket bar is a horizontal bar mounted about 18 inches (45 cm) off the floor that the bovine straddles. Brisket bars must never be used with chutes where the squeeze sides are hinged at the bottom because there will not be sufficient space for the cattle to walk. And if a complete squeeze chute is not available, a straight-bar stanchion head gate

Friction latches on a squeeze chute should never be oiled because that can cause the latch to fail to hold the head gate closed. This type of latch is quieter, but the disadvantage is that it must be carefully maintained to prevent accidental release.

is recommended, again to prevent choking. Cattle can usually lie down safely in a straight-bar stanchion.

Self-catchers are head gates equipped with two doors that close around an animal's neck when bumped by his head. They work best with gentle cattle and should not be used with wild horned cattle or bison. It is essential to adjust the self-catcher for cattle size. Severe injuries can occur if a self-catcher is adjusted too wide and the animal's shoulder passes partway through the closed gate. Cattle can also be injured if they run into the self-catcher at a high speed. Care must always be taken with self-catching head gates.

Manually operated squeeze chutes and restraints also require careful adjustment. These chutes have two types of latches: a friction latch (in which a rod moves through a hole in a spring-loaded metal plate) or a ratchet latch (in which a cog catches on a toothed ratchet). Friction latches have the advantage of being quieter and less disturbing to animals awaiting entrance, but they can be dangerous because they have a tendency to open unexpectedly when they start to wear out.

Latches and ratchet locks on squeeze chutes and restraints must be kept well maintained to prevent injury to people. If you notice that a ratchet device has become worn, replace it immediately. Though it's important to oil ratchet latches, friction-type latches must never be oiled. Oiling will destroy the ability of a friction latch to hold. On self-catching head gates, the mechanism must be maintained to prevent cattle from getting stuck partway through a closed gate.

Good Features for a Squeeze Chute or Tilt Table

- The floor is made of nonslip material.

- There are no pinch points.

- Both sides squeeze together simultaneously to keep the animal in a balanced position. The animal is more likely to become agitated if it is thrown off-balance.

- Manually operated chutes should be designed so that a person does not have to exert huge amounts of effort to operate them. Well-designed leverage systems require much less effort to operate. When buying a chute, try operating it at the dealer or on another farm.

- There is easy access to the neck so that injections can be given according to beef quality assurance guidelines. Injections in the rear of the animal damage the most expensive cuts of meat.

Deep-grooved concrete floors (fromGrandinLivestock HandlingSystems.com)

Troubleshooting Handling Problems

To solve a handling problem, determine the cause of the difficulty. Causes of handling problems can include one or more of the following:

- Scared, excited animals are given insufficient time to calm down. They need 20 to 30 minutes to settle.

- There is a facility design flaw such as a dead-ended chute (see chapter 5).

- The lighting is incorrect or a chute entrance is too dark.

- Small distractions (which can be easily corrected) cause balking.

- Too many cattle, pigs, goats, or bison are placed in a crowd pen. Fill the crowd pen only half full. This does not apply to sheep, who prefer to be bunched closer together. Moving sheep through a handling facility is like siphoning water: you never want to break the flow. All the other animals are moved in small separate bunches.

- Handlers get the animals agitated, excited, or scared.

- Genetics cause flighty, excitable animal temperament.

- Slick floors cause slipping and panic.

- Handlers on foot move livestock that have been handled exclusively on horseback. The cattle may panic and be difficult to handle the first time they are moved by a handler on foot.

- The animals have had a lack of exposure to people. Pigs, for example, differentiate between a person in an alley and a person walking through their pen. Pigs will be easier to drive if they have had experience with people walking quietly in their pens, which teaches them to quietly move away.

- Animals may refuse to enter a dark building on a sunny day. They will enter more easily if they can see through the building and see daylight on the other side.

- Handlers who stand in front of the point of balance and poke the rear confuse the animal, giving him simultaneous signals to go forward and backward.

You must determine whether you have a basic design problem, a small distraction that can be easily fixed, or an animal or handling technique problem before you begin work on the solution. If you zero in on the exact causes of the slowdown or stressed animal behavior, you'll avoid expensive, exhausting nonsolutions and save yourself time, energy, and money.

Using Visualization to Hold the Animal Gently

The hydraulic valves on a chute are like musical instruments: different brands have a different feel, just as different types of wind instruments do. Experimenting with operating the controls without any cattle in the chute enabled me to practice later via mental imagery. I had to visualize the actual controls on the chute and, in my imagination, watch my hands pushing the levers. I could feel in my mind how much force was needed to move the gates at different speeds. I rehearsed the procedure many times in my mind with different types of cattle entering the chute. When it was time to handle cattle, I was able to walk up to the chute and run it almost perfectly.

It worked best when I operated the hydraulic levers unconsciously, like using my legs for walking. If I thought about the levers, I got all mixed up and pushed them the wrong way. I had to force myself to relax and just allow the restrainer to become part of my body, while completely forgetting about the levers.

As each animal entered, I concentrated on moving the apparatus slowly and gently so as not to scare him. I watched the reactions so that I applied only enough pressure to hold him snugly. Excessive pressure would cause discomfort. If his ears were laid back against his head or he struggled, I knew I had squeezed him too hard. Animals are very sensitive to hydraulic equipment. They feel the smallest movement of the control levers.

Through the machine, I reached out and held the animal. When I held his head in the yoke, I imagined placing my hands on his forehead and under his chin and gently easing him into position. Body boundaries seemed to disappear, and I had no awareness of pushing the levers. The rear pusher gate and head yoke became an extension of my hands.

I need to add that before I could make this work, the cattle must be handled quietly when they are brought up to the restrainer. The day before, the handlers had learned how to move cattle calmly and no electric prods were used.

(From Temple Grandin, *Thinking in Pictures*, 1995)

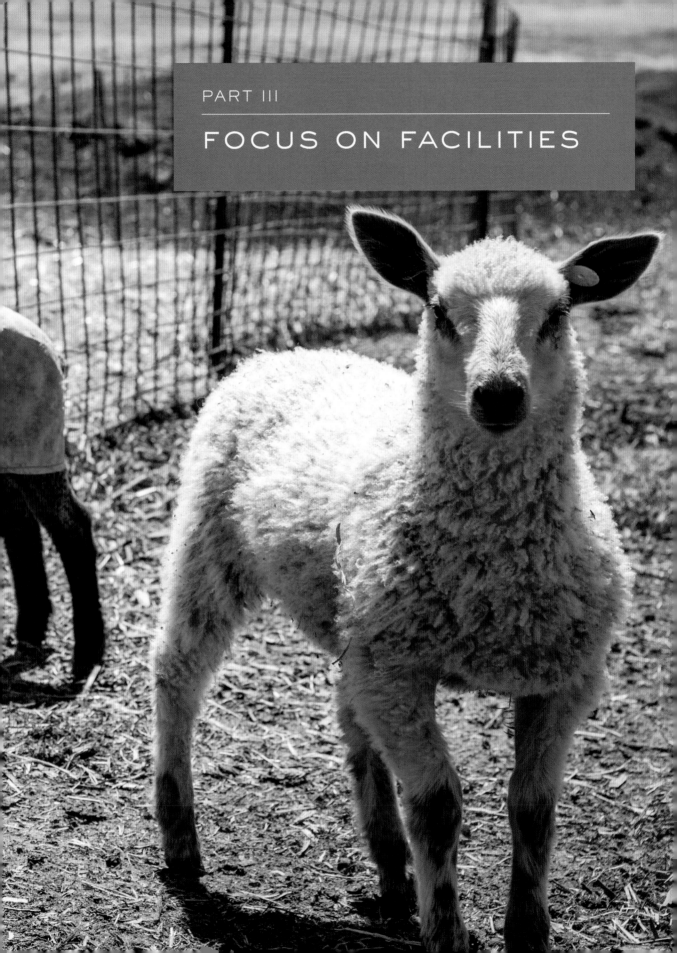

5

BUILDING OR BUYING HANDLING FACILITIES

When planning the design of your livestock handling facility, begin by establishing your needs and taking an inventory of your existing resources. If you own the land, you may opt to build a permanent facility with steel or wood posts firmly anchored in the ground with concrete. If you are leasing or renting land, you will probably opt to use manufactured handling equipment that can be set up as freestanding structures on top of the ground. In many states and municipalities, if you permanently anchor posts in the ground with concrete, you may not be allowed to remove the facility when you leave; it will be considered attached to the land. However, on your own land, you can build permanent facilities.

See my book *Humane Livestock Handling* (Storey Publishing, 2008) for details on construction methods for building chute systems from basic steel materials such as pipe and metal sheets. The book also has designs for easy-to-make hinges, latches, and slide gates. If you supply the labor, the costs of permanent

PRINCIPLES

Correct layout is essential for efficient livestock movement.

There are many premanufactured livestock handling systems, but most make the same basic parts.

Different types of handling facility designs have advantages and disadvantages.

Floors in scales, squeeze chutes, chutes, and crowd pens must be nonslip. Animals become agitated and may panic on a slippery floor.

facilities may be lower than those for portable ones. Welding and building skills will be required. If you do not know how to weld, you may want to take classes at a local community college or technical school. Your other alternative is to construct your facility from manufactured components.

If you plan to build on land with no existing structures, you are beginning with a clean slate and your planning stages will be consumed with considerations of cost, site location, and water drainage. If your goal is renovation, you need to design your facility to fit around existing structures that cannot be removed.

Animal behavior considerations do not end when you design your system. It is best to make a scale drawing of what you plan to build. When you design it to scale, you will avoid many mistakes. Even if you are using manufactured components, it is still recommended to make a scale drawing. Then you can figure out exactly which components you will need to purchase. There are two simple items you will need to do scale drawings:

A. **Graph paper.** One square on the graph paper is equal to either 1 foot or 1 meter. Use paper with larger squares for meters.

B. **Scale rulers.** These rulers are available in either English or metric measures. For English measurements, use an engineering scale where 1 inch equals 10 feet. For metric measurements, use a metric scale ruler that can measure whole meters and fractions of a meter.

Throughout the planning stages, you must keep in mind the animals' natural following behavior, their desire to return to where they came from, and their anxiety when their flight zone is deeply invaded and they are not able to move away.

New Facilities

It is essential that there is sufficient land available on the site you select. Choose a location that is not wet or boggy, or you will be working in a mud pit, which is stressful for both animals and handlers. Don't build over the tops of septic tanks, and make sure the site drains well, with no runoff into streams. Access to electrical power and water is preferable. Be sure to obey local building ordinances, and pay particular attention to property-line buffers and rights-of-way onto state and federal highways.

Make one of your earliest pieces of footwork a trip to the town hall to learn about building codes and permits. A building permit for a corral is not required in most states. However, if you erect a building over your facility, a building permit may be required. For portable manufactured facilities that sit on top of the ground, no permits are required. Another factor you should consider is depreciation for tax purposes. Portable facilities can be depreciated much quicker than buildings or permanent installations. You will need to discuss your options with a tax accountant.

Some handling facility designs are specifically designed for cattle, and others for sheep and goats. Designs for sheep will also work well for goats. Many producers have both cattle and sheep or goats on the same farm. Round crowd pen cattle layouts also can be used for sheep and goats. It is strongly recommended to construct an additional small single-file chute to handle the smaller animals. Handling sheep or goats in a single-file chute designed for large cattle will be difficult because they will turn around. You will need a gathering pen that can hold the entire herd or flock. Due to their horns and independent nature, goats are more likely to get stuck in fences compared to sheep. Fences that are adequate for sheep may have openings that goats may get hung up on.

FIGURE OUT SPACE FOR DIFFERENT HANDLING FUNCTIONS

Make a list of all the procedures you want to perform in the handling facility. Once you've completed your plan, be sure that your drawing has accommodated pen space for all of them.

Plans for a new facility for all livestock should include a gathering pen, a handling facility, and a post-working pen to hold animals as they exit the squeeze chute or handling race. The gathering pen site should be located where animals can be easily herded in from the pasture. This pen is used to hold animals before sorting or working them in the handling facility.

Calculate the space required for gathering all your animals. For example, a pen 35 feet (10.6 m) wide by 25 feet (7.5 m) long should be able to hold 20 mother cow/calf pairs (see box below). These space recommendations also provide sufficient "working" space. If a pen is too small and animals are jammed in tightly, it is much more difficult to quietly move them in or out of the pen. Provide additional space for animals to lie down if they have to spend the night in the pen.

Space Requirements for Holding Animals in a Handling Facility

CATTLE

Cow and calf pairs: 35 to 45 square feet (3.25 to 4.18 m²) per pair, depending on calf size

Cows with calves over 350 pounds (163 kg): more than 35 square feet per pair

Adult cows and market-weight fed cattle: 20 square feet (1.86 m²). Cattle weighing over 1,400 lbs (635 kg) will require 22 sq ft (2 m²).

SHEEP and GOATS

Large-breed ewes with wool: 6.5 square feet (0.6 m²)

Large-breed lambs and goats: 5 square feet (0.47 m²)

A large woolly ewe with a large lamb: 10 square feet (0.92 m²)

A small woolly ewe: 5 square feet (0.47 m²)

SWINE

Market pigs: 5 to 6 square feet (0.47 m² to 0.55 m²), depending on size. Extremely large pigs over 280 pounds (127 kg) will require more space.

BISON

Bison require more space than cattle because they may fight and get seriously injured in a crowded holding pen.

WHAT OLD STUFF DO YOU PLAN TO KEEP?

Certain structures are too expensive to move. Barns, shops, power poles with transformers, water wells, livestock waterers set on concrete slabs, and paved roads are all examples of permanent structures that usually have to be incorporated into planning and design. Draw all these "site restrictions" on a scale drawing so that you can design your handling facility around them. Existing structures that often should be removed are wire fences, old broken corrals, a dilapidated barn, or a crumbling feed bunk.

Don't let a single existing fence ruin a good plan. Many people try to keep too many old fences. Include on your planning sheet *only* the dimensions of the land and the structures that cannot be moved or torn down. That way, you can draw a good design without your mind being boxed in by old fences and other stuff you will be removing.

Design Concepts of Livestock Handling Systems

There are two different concepts to livestock handling facility design: build a more elaborate facility that requires less skill to use effectively and safely, or build a very simple economical design that requires greater stockmanship skill.

When a single-file chute with open sides is used, the skilled stockperson remains outside the animal's flight zone except when he or she enters the flight zone to move an animal. After the animal is moved, the handler backs away from the open side to get out of the flight zone. If the handler continually stands inside the flight zone, the animals may become agitated or restless.

If the animals are really wild and not accustomed to being handled, the use of solid sides is strongly recommended. Solid sides are usually recommended on outside perimeter fences to block distractions, such as vehicles, farm equipment, or people walking by. The skilled stockperson works along inner fences that are open. The place where totally open perimeter fences work best are on remote pastures with no outside distractions.

THE IMPORTANCE OF NONSLIP FOOTING FOR ANIMALS

All species of animals become agitated and frightened when they slip. Since many small producers will be using pre-manufactured facilities, the floor may be natural dirt. This will provide a good nonslip surface in areas with low rainfall. In areas with high rainfall where deep mud is a problem, it is strongly recommended to have a concrete floor in facilities that will remain in a permanent location. The concrete should be placed under the squeeze chute and the single-file chute or double-file chute. See my book *Humane Livestock Handling* for more information on grooving concrete for a nonslip walking surface. A smooth finish or a rough broom finish will not work. Animals will slip and fall. Deeper grooves are essential. For facilities that will be frequently moved, concrete is usually not required.

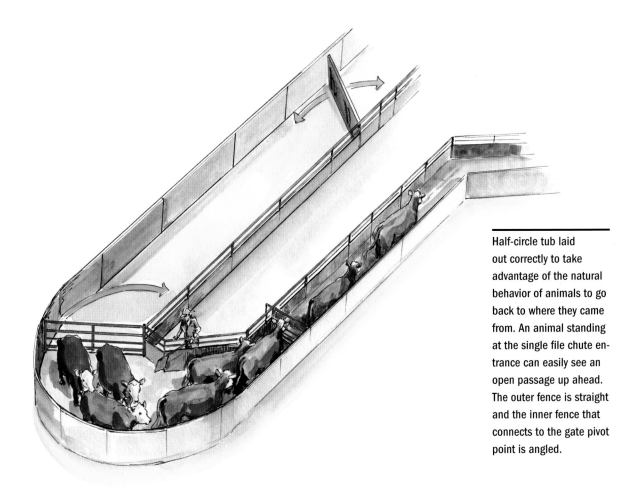

Half-circle tub laid out correctly to take advantage of the natural behavior of animals to go back to where they came from. An animal standing at the single file chute entrance can easily see an open passage up ahead. The outer fence is straight and the inner fence that connects to the gate pivot point is angled.

The illustrations on the next several pages show the basic curved layouts for moving cattle into a squeeze chute. A correctly designed and constructed curved system is efficient for several reasons. Grazing animals have a tendency to want to go back to where they came from. This is a natural instinctual behavior. Designs that take advantage of that behavior will facilitate animal movement.

AVOID LAYOUT MISTAKES

Layout mistakes can ruin a handling system. All handling system designs must be laid out correctly to work efficiently.

Below are a number of design recommendations. The correct layout for a tub system is shown above.

Never dead-end the entrance of the single-file chute. An animal standing at the entrance of a single-file chute must be able to see two body lengths up the chute. The most critical part of the layout is the junction between the single-file chute and the crowd pen. There should be a minimum of a 10-foot (3 m) straight section before the chute angles or curves. Dead-ending the chute, one of the most common layout mistakes, will ruin a system.

BAD DESIGN: CHUTE IS BENT TOO SHARPLY

This single-file chute is completely dead-ended.
The dotted lines show how to correct this common
layout mistake.

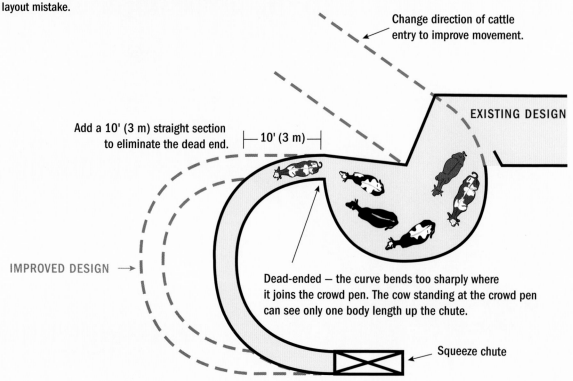

Change direction of cattle
entry to improve movement.

EXISTING DESIGN

Add a 10' (3 m) straight section
to eliminate the dead end.

|— 10' (3 m) —|

IMPROVED DESIGN →

Dead-ended — the curve bends too sharply where
it joins the crowd pen. The cow standing at the crowd pen
can see only one body length up the chute.

Squeeze chute

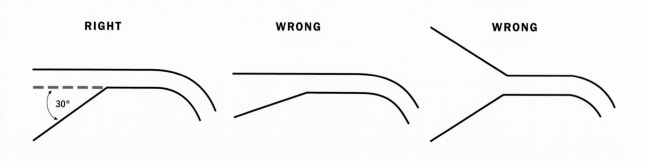

| RIGHT | WRONG | WRONG |

30°

This illustration shows both correct and incorrect layouts of a crowd pen entrance angle for cattle, sheep, goats, and bison.

Use the correct angle on the crowd pen.
In a crowd pen for cattle, sheep, goats, or bison, one side should be straight and the other side should be on a 30-degree angle. The handler should work from the angled side. Do not, however, use this design for pigs. Pigs will jam in a funnel. A pig crowd pen should have a single offset step equal to the width of one market-weight pig.

A round crowd pen (tub) should make a 180-degree half-circle turn. The direction of travel through this half-circle takes advantage of the natural behavior of all species of animals to want to return to where they came from, where they last felt safe. Crowd pens work best for cattle and pigs when they are used as "passing-through" pens. Cattle, goats, and pigs should be handled in small, separate bunches. Sheep have such intense following behavior that they can be handled in large groups that continuously flow through the facility.

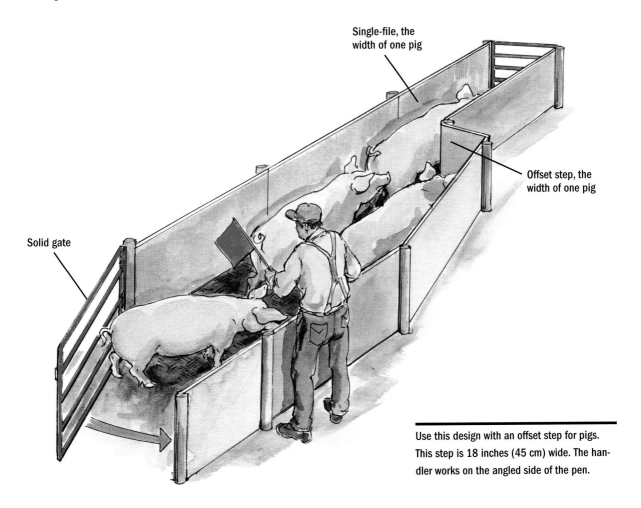

Single-file, the width of one pig

Offset step, the width of one pig

Solid gate

Use this design with an offset step for pigs. This step is 18 inches (45 cm) wide. The handler works on the angled side of the pen.

SORTING COWS AND CALVES

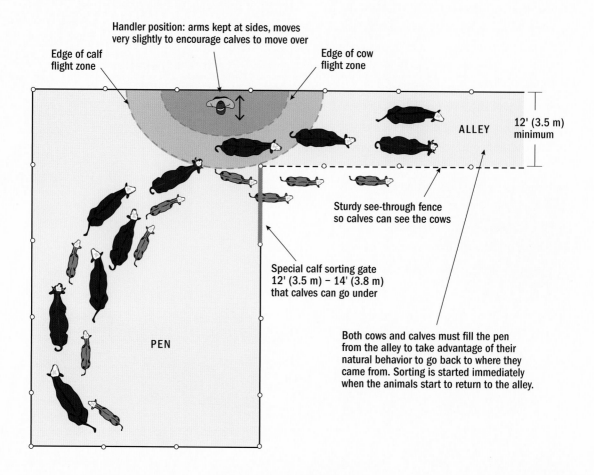

Handler position: arms kept at sides, moves very slightly to encourage calves to move over

Edge of calf flight zone

Edge of cow flight zone

ALLEY

12' (3.5 m) minimum

Sturdy see-through fence so calves can see the cows

Special calf sorting gate 12' (3.5 m) – 14' (3.8 m) that calves can go under

PEN

Both cows and calves must fill the pen from the alley to take advantage of their natural behavior to go back to where they came from. Sorting is started immediately when the animals start to return to the alley.

Calf and cow sorting system designed by Joe Stookey, University of Saskatchewan. Calves tend to move farther away from the handler than cows because they have a larger flight zone. To stay out of the "bubble" formed around the handler, the calves escape through the sort gate.

Basic Components of Manufactured Animal Handling Systems

Manufactured systems consist of different components. There are many different manufacturers, but they make many of the same basic parts. The next section describes the different parts you can purchase.

Squeeze chutes or restrainers for cattle, bison, sheep, or goats. Many companies make different types of squeeze chutes, hoof-trimming tables, and other equipment. For all types of animals, the best tilt-table systems fully support the animal and do not allow it to flop around (see chapter 4). For beef calves, the systems that are most comfortable for the animal consist of a miniature squeeze chute that tilts the calf on its side. Squeeze chutes

for cattle or bison and other restrainers for goats or sheep are purchased as complete units, with minimal or no assembly required.

Fence panels. Fence panels are available in different heights for cattle, sheep, or goats: 6 feet (2 m), 8 feet (2.5 m), 10 feet (3 m), 12 feet (3.5 m), 14 feet (4.2 m), and 16 feet (5 m). Strong, extra-tall fence panels are available for bison. The most commonly used cattle panel is 12 feet (3.5 m) long; longer panels are harder to pick up and move.

Gate panels. Gate panels come in the same lengths as fence panels and have a frame in which the gate swings. The frame enables the gate to easily swing and latch.

A creep feed gate allows young lambs or calves to access feed, but the cows and ewes will not fit through the openings. Note that all the sheep have at least one ear facing the photographer, while the two that have already passed through are looking at him. Animals want to see you.

Walk-through panels. These have a small gate in a fence panel for people to walk through. In cattle facilities, walk-through panels should be located where indicated on the diagrams for safety; they should be placed so that handlers do not have to climb fences.

Lead-up alley (chute, race) components. The following components are used to guide animals in a single file:

A. Straight, adjustable sections of single-file chute for different-sized cattle, sheep, or goats

B. Curved sections of single-file chute to form complete systems

C. Angled sections of single-file chute with 45-degree bends

D. Complete double-file alley (chute) systems

Tubs and round crowd pens. These enclosures are available in quarter-circle and half-circle sections. Most commercial portable tubs for cattle have a radius of 10 feet (3 m). Do not use a shorter radius for cattle. A manufactured sheep or goat system usually has a radius of 8 feet (2.5 m).

Sliding gates. Used at the entrance to the single-file chute for all species. For cattle and bison, sliding gates move horizontally so that they are easy to operate manually. Some sheep and goat facilities have vertical slide gates.

Backstops. Small gates that prevent animals from backing up. Available for all species.

Portable loading ramps. Ramps are used for loading large trucks. Most small producers will use a stock trailer and do not require a loading ramp.

An agitated, fearful calf switches his tail in the squeeze chute. Tail switching often occurs before the animal becomes agitated.

Types of Portable Trailer-Mounted Handling Systems

Trailer-mounted handling systems are recommended when animals have to be handled in many different pastures. Local livestock associations can buy a system that many producers can use. Such systems may consist of the following components:

Trailer with portable truck-loading ramp and fence panels. This consists of a single trailer-mounted loading ramp. There are units available that can also transport a bunch of fence panels. This unit is ideal for moving animals between pastures. The panels are used to make a corral leading to the loading ramp.

Trailer for transporting panels. One trailer can transport up to 20 cattle panels or an entire sheep or goat handling facility.

Single trailer with complete cattle handling facility. The complete facility consists of a squeeze chute, a lead-up alley, and a crowd pen. Some of these units have racks for fence panels, and some do not. The advantage of this unit is that the squeeze, alley, and crowd pen are all on a single trailer. This is one reason why these units are popular. The disadvantage of some of these units is that they may not have sufficient single-file chute length to encourage following behavior.

Folding portable corral trailer. This trailer has panels on wheels so that fence panels

do not have to be carried to set it up. It works really well for loading stock trailers.

Trailer-mounted squeeze chute. This should be used with other components to form a complete cattle handling facility.

Trailer-mounted squeeze chute and lead-up alley. These units provide more lead-up chute capacity than a trailer that also includes the crowd pen. Trailer-mounted systems are available in either a single- or double-file alley, and some are designed to also carry a number of fence panels. This is an application where a Bud Box (see page 141) is recommended, because it can be set up easily with the portable fence panels. This is more convenient when the facility has to be frequently moved.

Building Placement to Facilitate Animal Movement

How a building is placed over a handling facility can have a big effect on ease of animal entry. On a bright sunny day, animals may refuse to enter a dark building. I call this the "black hole" or dark movie theater effect. Illumination with lamps will attract animals in at night, but will not work on sunny days. If approaching animals can see daylight through the other side of the building, this will often facilitate entry. If you have an existing "black hole" problem, putting up shade cloth or other shade over the alley entering the building may facilitate entry because it reduces contrast between light and dark. If shade cloth is used, it must not flap.

Place building in the correct spot. Never place the wall of a building or the edge of a roof at the junction between the single- or double-file chute and the crowd pen for any species. The contrast in lighting (moving from the bright outside to the inside of a dark building) at this critical juncture will cause animals to balk or turn back. Either a building should cover the entire facility or just a small building should be used over the squeeze chute.

Entire handling systems for small-scale sheep or goat flocks can fit easily inside a building. For cattle or bison, however, smaller buildings are preferred to large buildings because they can be placed just over the squeeze chute, can be easily heated in the winter, and are easier to keep clean. Large buildings tend to become homes to birds and other pests nesting in the rafters and elsewhere. For cattle, a minimum of 20 feet (6 m) of single-file chute should extend outside the building.

Animals readily enter buildings if abundant daylight enters with them. Large enclosed buildings should have white translucent skylights. For all species, animals enter more easily if they can see daylight through a building. All species often enter a building more easily when it has either open sides or a large open door that allows approaching animals to see through to the opposite wall.

Cattle will move easily into a building where the squeeze chute is located because they are lined up in a single file *before* entering the building. On a windy day, the plastic strip curtain may cause animals to balk.

Don't build overhead catwalks. Handlers should be either on the ground or on a low catwalk alongside the fences. All species become agitated and confused when people are above them. The designs in this book for small producers will not need catwalks, unless the producer opts for totally solid sides.

Build the correct crowd pen entrance angle. All the layouts in this book for cattle, bison, and sheep have a crowd pen entrance with one straight side and one side on a 30-degree angle (see page 132). This funnel design works poorly with pigs. The junction between a crowd pen and a single-file chute for pigs must be abrupt to prevent jamming. Pigs have a tendency to push forward when they panic.

Level ground is recommended. Animals move easily if the crowd-pen-and-chute system is level. Except for a slight drainage slope, all ground should be level. If you must build on a sloped site, the chutes and crowd pen should be on the highest ground so that they remain dry.

Consider bright sunlight. Do not face the squeeze chute, sheep restrainer, or loading ramp into the rising or setting sun. Animals balk and refuse to move into blinding light.

Avoid too many backstops. If animals are constantly backing up, you need to figure out what is causing it instead of installing more backstops. People often install too many backstops. Many times I have improved animal movement by tying backstops open.

Decisions about Solid Sides

Some advocates of low-stress cattle handling believe that all solid sides should be removed from single-file chutes and crowd pens. This enables the cattle to easily see a person on the ground and calmly move in response to the handler's movements. For this approach to work, the handler must have a very high level of skill and the facility must be free of all the distractions listed in the table on page 115. Remove solid sides only if the cattle will not see dogs, vehicles, windmills, people (including children) standing in the wrong places, and other distractions outside the fence. Open-sided facilities work best on remote pastures.

The curved structures with completely solid sides shown in this book achieve safe, low-stress handling with relatively little skill. Crowd pens and chutes with open fences require more skill and are thus most likely to be effective when used by motivated people who are willing to learn more advanced techniques. Included in this section are three diagrams of facilities where solid sides have been partially removed so that a handler on the ground can move cattle by using the principles of flight zone and point of balance. The remaining solid sides are on the outside perimeter of the facility to block distractions such as parked vehicles or people walking by.

Figure 1 shows a standard curved layout with solid sides, with a small open-sided 7-foot (2 m) section of single-file chute at the squeeze chute entrance. To prevent leg injuries, the bottom 3 feet (1 m) of the fence is solid. To move a cow into the squeeze chute, a person working at the squeeze walks back just past the animal's eyes.

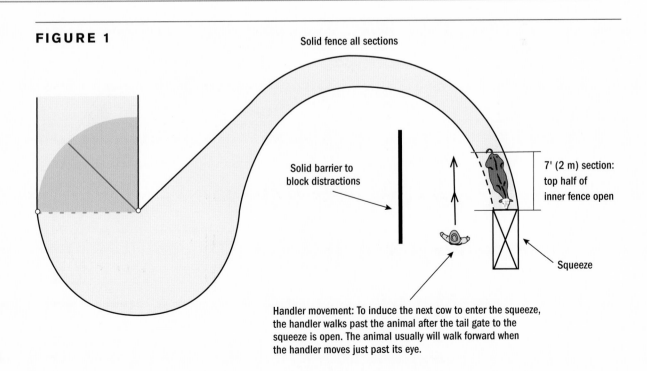

FIGURE 1

Solid fence all sections

Solid barrier to block distractions

7' (2 m) section: top half of inner fence open

Squeeze

Handler movement: To induce the next cow to enter the squeeze, the handler walks past the animal after the tail gate to the squeeze is open. The animal usually will walk forward when the handler moves just past its eye.

The movement has to be done at a fast walk so that the animal anticipates that the handler will pass the point of balance at the shoulder. If the movement is too slow, the cow will back up. It is easier to move cattle into the squeeze chute if the back half of the chute is covered so that the animals do not see the operator standing next to the squeeze as they enter (see page 111). Also shown in this figure is a solid barrier to prevent a cow that is waiting at the squeeze chute entrance from seeing people in the crowd pen area.

Figure 2 shows a modification of the basic curved design that has worked really well in Australia. All of the fences on the outer perimeter of the single-file chute are solid to block distractions, and the round crowd pen also has a completely solid fence. However, the top of the inner fence of the curved single-file chute is open. As a result, people *must* stay out of the inner radius area, except when it is entered to move cattle. It is strongly recommended that the bottom 3 feet (1 m) of the open fence be solid to prevent legs from becoming caught and injured.

These open-side designs should not be used in slaughter plants or other areas where there is a high number of distractions or untrained people who walk in all the wrong places. Truck loading ramps should have completely solid sides, because there are often lots of outside distractions or high numbers of poorly trained people who may be using them.

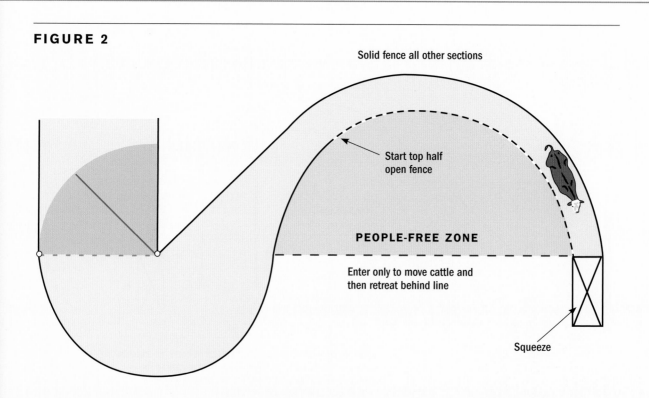

FIGURE 2

Solid fence all other sections

Start top half
open fence

PEOPLE-FREE ZONE

Enter only to move cattle and
then retreat behind line

Squeeze

Bud Box Cattle Return Alley

Figure 3 shows a Bud Box, a simple and very economical design developed by cattle handling expert Bud Williams that can be set up easily with portable panels. It takes advantage of the natural tendency for cattle to go back to where they came from. Another name for this design is a return alley. Do not use this design for sheep or goats, because it may increase the chances of animals piling.

The cattle enter the Bud Box and turn around and head back toward the entrance gate. A handler positioned in the Bud Box guides them into the single-file chute. This design can also be used with a straight double-file chute (see page 103). The Bud Box works best where there is sufficient space in the single-file or straight double-file chute to hold all of the cattle that are put in the Bud Box. To maximize following behavior, the lead-up chute should be nearly empty when the Bud Box is filled. This enables the cattle to turn back and immediately start filling the chute leading to the squeeze.

Only people who have achieved a high level of stockmanship skill should use the Bud Box because the handler is inside a confined area with the cattle. An unskilled person is more likely to be injured in a Bud Box than in a crowd pen where the handler works outside the fence. The box works best on places where a few highly skilled people use it or in operations where a person on a horse is inside the Bud Box. Another alternative is to work the Bud Box from outside the fence with a long flag. Cattle must be moved in very small groups of 8 cows or 10 calves because the Bud Box will not work if it is more than half full. Crews that work cattle in multiple locations with both round tubs and Bud Boxes report that it takes a longer time to train a new employee to use a Bud Box correctly. A new employee can usually be trained to use a round tub correctly in a single day.

The back gate of the Bud Box should be completely solid to help prevent cattle from turning back on the handler. To block distractions, the outer perimeter fences should be solid and the inner fences where the handler works should be open. In places with lots of distractions, such as feedlots, place an additional solid barrier or a building over the facility. Some of the most effective Bud Box systems are inside a large building with solid sidewalls to block distractions that occur along the perimeter. The far end of the Bud Box should be open to encourage the cattle to enter. Cattle will not enter a place that looks like a dead-end. The Bud Box should

FIGURE 3

Bud Box Return Alley should be used only by very experienced stock handlers.

– – – – – – OPEN FENCE

———— SOLID FENCE

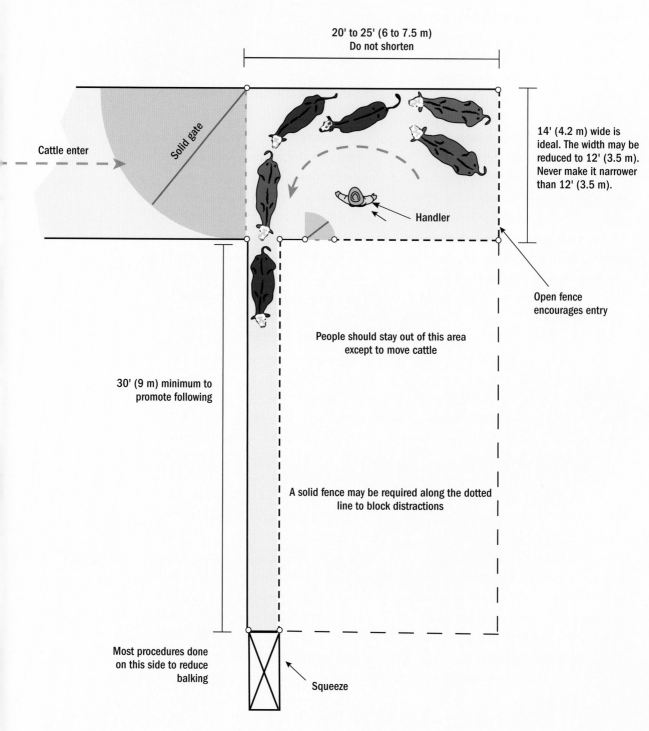

20' to 25' (6 to 7.5 m)
Do not shorten

Cattle enter

Solid gate

14' (4.2 m) wide is ideal. The width may be reduced to 12' (3.5 m). Never make it narrower than 12' (3.5 m).

Handler

Open fence encourages entry

People should stay out of this area except to move cattle

30' (9 m) minimum to promote following

A solid fence may be required along the dotted line to block distractions

Most procedures done on this side to reduce balking

Squeeze

be equipped with a small mangate so that the handler can easily escape.

The open end must be free of distractions such as vehicles or pedestrians. Distractions seen through the open end may cause cattle to turn back before the handler can close the entrance gate. For handler safety, dogs must be kept completely away from the Bud Box. This system must be laid out correctly. The minimum width is 12 feet (3.5 m) — do not make narrower. Bud Boxes used exclusively with horses should be 14 ft (4.25 m) wide. This design is always in the shape of a rectangle. It should always be longer than its width. The single-file chute must be laid out at a 90-degree angle in relation to the Bud Box. If a curved single-file chute is used, there *must* be a 12-foot (3.5 m) straight section where the chute joins the Bud Box. If a single- or double-file chute is curved or angled where it joins the Bud Box, the system may be dead-ended and the cattle may refuse to enter the chutes.

Working the Gate Pivot Point in a Round Crowd Pen (Tub)

The layout on the next page and the one shown on page 145 are both designed so that all expensive catwalks are eliminated. The outer perimeter fences are solid, and the handler works in the inner areas and walks on the ground. The inner fences should be solid only on the bottom portion.

After the handler has filled the crowd pen and positioned the crowd gate as shown on the diagram, he or she then moves to the gate pivot point. This uses the same principle as the Bud Box, but safety for the handler is improved. The handler uses the flag to quietly urge the cattle to circle around and enter either a single-file or double-file chute. They will move easily into the chute because they are going back to where they came from. The only time the crowd gate may need to be closed more tightly is when a single agitated animal gets left behind.

The figure on page 145 illustrates a layout where the handler has easy access to both the gate pivot point and the inner radius of the curved single-file chute. Both of these layouts use the funnel layout shown at the top of page 126 and on page 131. A small step installed at the pivot point will make it easier for the handler to reach over the fence with the flag.

HALF-CIRCLE TUB
In this layout the handler works at the gate pivot point.

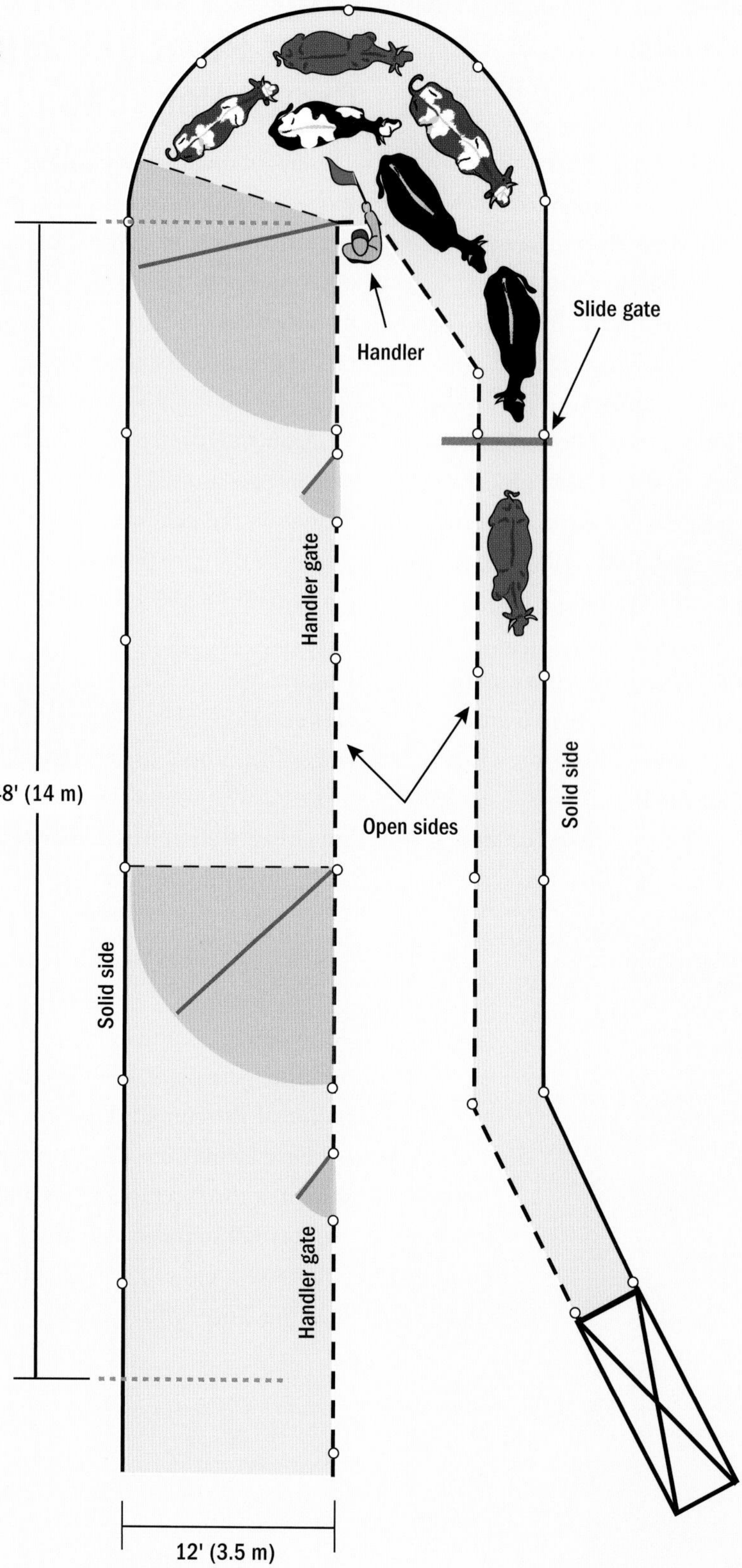

Handler

Slide gate

Handler gate

Open sides

Solid side

48' (14 m)

Solid side

Handler gate

12' (3.5 m)

QUARTER-CIRCLE TUB AND CURVED ALLEY

This layout takes advantage of the natural behavior of cattle to go back to where they came from.

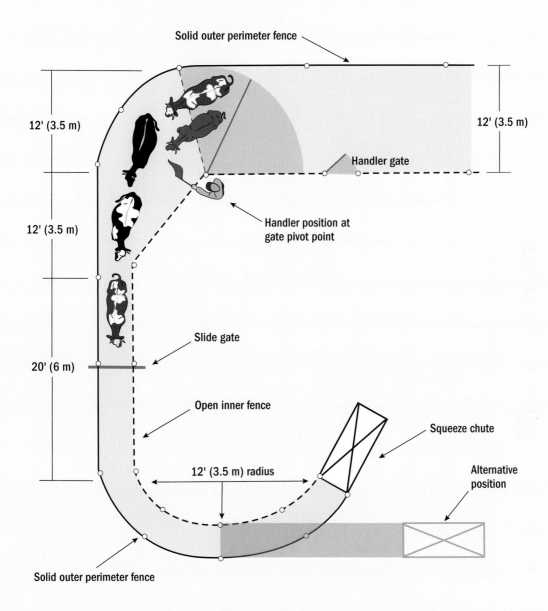

Solid outer perimeter fence

12' (3.5 m)

12' (3.5 m)

Handler gate

Handler position at
gate pivot point

12' (3.5 m)

Slide gate

20' (6 m)

Open inner fence

Squeeze chute

Alternative
position

12' (3.5 m) radius

Solid outer perimeter fence

6 | CORRAL LAYOUTS

I have designed 11 new working facility layouts that can be easily constructed from portable panels and other premanufactured components. All the designs take advantage of the behavioral principles of animals wanting to go back to where they came from. The entrance to the single-file race or double-file race can be equipped with either a sliding entrance gate or one-way backstop gates. Animals usually move more easily through a facility if they do not have to push up one-way backstops and go through them. A common mistake is to install too many backstops or make the pivot points too low for tall animals.

Some commercially available one-way backstop gates are designed so that they can be opened easily and then held open. This type is strongly recommended. Some backstops are hinged on the side of the single-file chute. Backstops that are hinged on the top are preferable because they are usually easier to hold open and then close after cattle or sheep pass through. A backstop that pivots on the top can be easily rigged with a remote control rope. This enables a person who is located in the crowd pen area to hold the gate open for the animals. Sliding gates are more expensive than backstops, but cattle often will pass through them more easily.

PRINCIPLES

The best chute and corral designs use the behavioral principle of cattle, sheep, and other livestock wanting to go back to where they came from.

Round tub systems take advantage of the natural tendency of cattle to circle around the stockperson.

Bud Box systems are simple and economical, and work well for cattle.

Simple, efficient designs for sheep and goats utilize full- or half-circle round tubs.

Adjusting the Size of Single-File Chutes That Lead to the Squeeze

Many commercially available single-file or double-file chutes are designed so that they can be adjusted for different-sized cattle. When shopping for chute sections, look for ones that are easy to adjust. Straight sections are often easier to adjust than curved sections. Another option is to have two chutes with stationary sides, one for adult cows and another for calves or sheep.

There have been many discussions about straight chutes versus V-shaped single-file chutes. A stationary V-shaped chute can handle a variety of cattle sizes with no adjustment. The disadvantage is that large cows from breeds that have really wide bodies may have difficulty walking through them. V-shaped chutes work well for younger cattle handled in feedlots.

Approximate Metric Conversions

Many countries have standard metric sizes for gates and panels, but they can vary from country to country. The following diagrams have absolute conversions. Below are what I call "rounded" metric conversions to make it easier to do the math and make estimates.

8 ft = 2.5 m
10 ft = 3 m
12 ft = 3.5 m

Simple Facility for 5 to 10 Cows

The dotted line shows the pathway of the cattle through the facility. The pasture entrance is positioned to take advantage of the natural behavior of cattle to return to where they came from. The handler works the animal's flight zone and point of balance along the fences of the inside area. The outer fences are covered if there are distractions outside the facility such as vehicles and people walking by. A handler working in the inner area takes advantage of the animal's tendency to circle around him or her. To load stock trailers, the trailer is backed up to the entrance of the holding alley. All catwalks are eliminated, and the inner fences are open on the top so that the handler can remain on the ground. This layout can be easily built from portable fence panels.

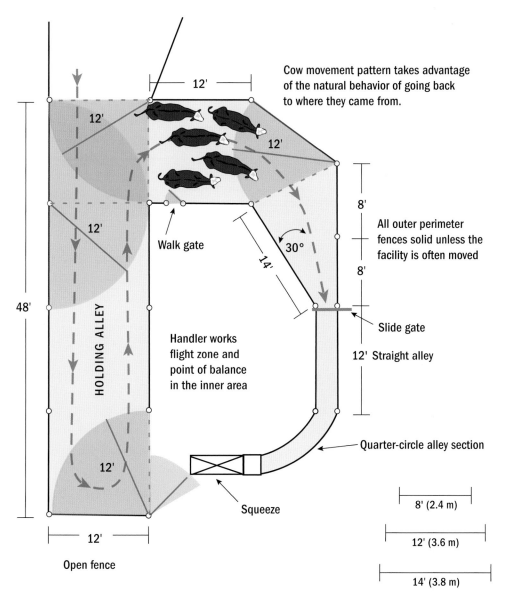

Cow movement pattern takes advantage of the natural behavior of going back to where they came from.

All outer perimeter fences solid unless the facility is often moved

Walk gate

30°

14'

Slide gate

12' Straight alley

HOLDING ALLEY

Handler works flight zone and point of balance in the inner area

Quarter-circle alley section

12'

Squeeze

Open fence

8' (2.4 m)

12' (3.6 m)

14' (3.8 m)

48'

12'

12'

12'

12'

12'

12'

8'

8'

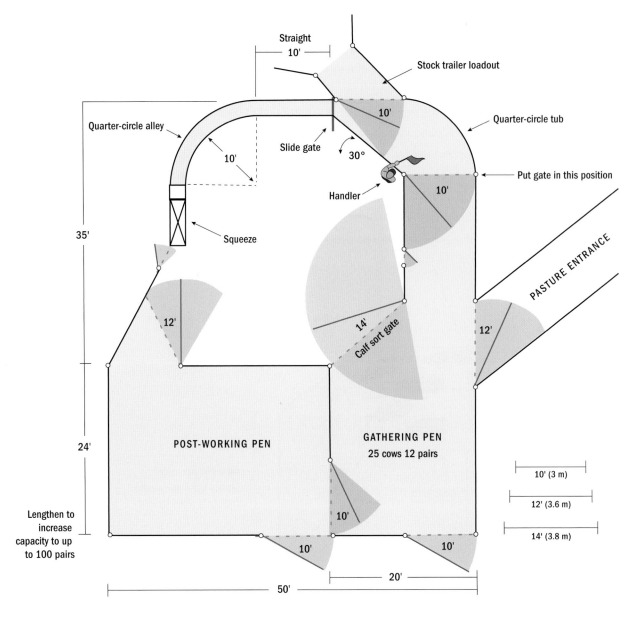

Straight
10'

Stock trailer loadout

Quarter-circle alley

10'

10'

Quarter-circle tub

Slide gate

30°

Put gate in this position

Handler

10'

35'

Squeeze

PASTURE ENTRANCE

12'

12'

14'

Calf sort gate

24'

POST-WORKING PEN

GATHERING PEN
25 cows 12 pairs

10' (3 m)

12' (3.6 m)

Lengthen to
increase
capacity to up
to 100 pairs

10'

14' (3.8 m)

10'

10'

20'

50'

Curved Facility for 25 Cows or 12 Cow/Calf Pairs

All movements of the handler are inside the inner area to take advantage of the animal's tendency to circle around him or her. The handler works from the center pivot of the crowd gate to load stock trailers or move cattle into the squeeze chute. The outer perimeter fences are totally solid, and the inner fences are open on the top, so the handlers can work on the ground. The pasture entrance is located to take advantage of the natural tendency of cattle to go back to where they came from. This will make it easier to move cattle into the quarter-circle tub. This layout also can use the calf sorting gate, which is described in chapter 5 (page 134). Another good feature is a 12-foot (3.5 m) sort gate in front of the squeeze chute. If an animal is mis-caught in the squeeze chute, he can be easily sorted out. Expansion of this design can be easily done by expanding the gathering and post-working pens.

Double Alley and Half-Circle Tub

This layout is set up for 35 cows or 15 cow/calf pairs. The handler works the center pivot point of the crowd pen to take advantage of the natural tendency of cattle to circle around the handler. It works on the same principle as the Bud Box because the cattle circle around the handler. The crowd gate is put on the first notch. A commercially available double alley and half-circle tub are used. The gathering pen and post-working pen can be easily constructed from 12-foot (3.5 m) panels. Filling the tub will be easy because the pasture entrance location takes advantage of the natural tendency of cattle to go back to where they came from. The system can be easily expanded by enlarging the gathering and post-working pen.

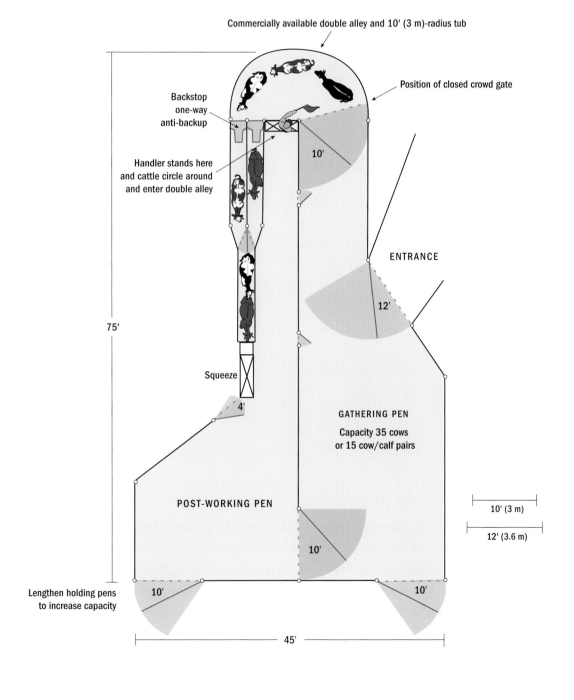

Commercially available double alley and 10' (3 m)-radius tub

Position of closed crowd gate

Backstop one-way anti-backup

Handler stands here and cattle circle around and enter double alley

10'

ENTRANCE

12'

75'

Squeeze

4'

GATHERING PEN

Capacity 35 cows or 15 cow/calf pairs

10' (3 m)

12' (3.6 m)

POST-WORKING PEN

10'

Lengthen holding pens to increase capacity

10'

10'

45'

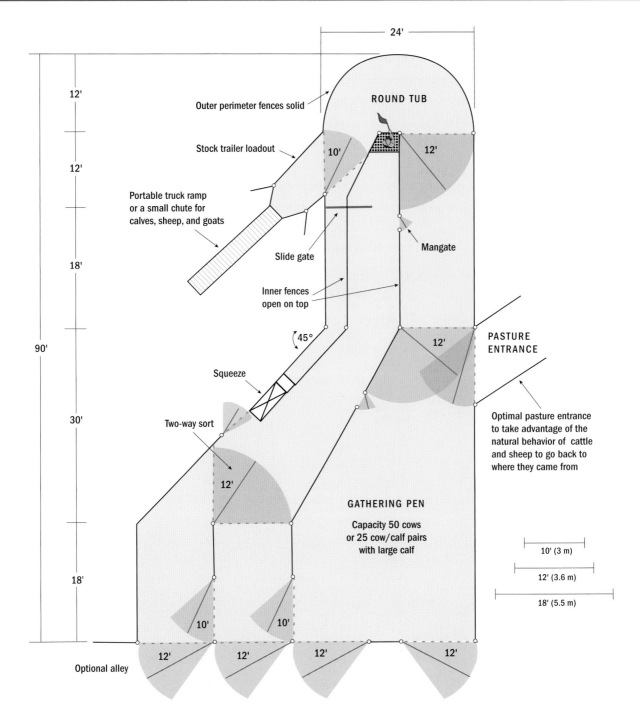

24'

12'

12'

18'

90'

30'

18'

ROUND TUB

Outer perimeter fences solid

Stock trailer loadout

10'

12'

Portable truck ramp
or a small chute for
calves, sheep, and goats

Slide gate

Mangate

Inner fences
open on top

45°

**PASTURE
ENTRANCE**

12'

Squeeze

Two-way sort

12'

Optimal pasture entrance
to take advantage of the
natural behavior of cattle
and sheep to go back to
where they came from

GATHERING PEN

Capacity 50 cows
or 25 cow/calf pairs
with large calf

10' (3 m)

12' (3.6 m)

18' (5.5 m)

10'

10'

12'

12'

12'

12'

Optional alley

Half-Circle Tub with Two Branches

This layout, set up for 50 cows or 25 cow/calf pairs, is very similar to the preceding figure and uses the same principles. The second branch can be used for loading stock trailers, or a separate small chute can be added for either calves, sheep, or goats. Capacity and cost can be easily reduced by making the gathering pen and the post-working pen smaller. The outer perimeter fences are totally solid, and the inner fences are open on the top. A pasture entrance can also be installed in the same location as on page 151 to facilitate filling the crowd pen.

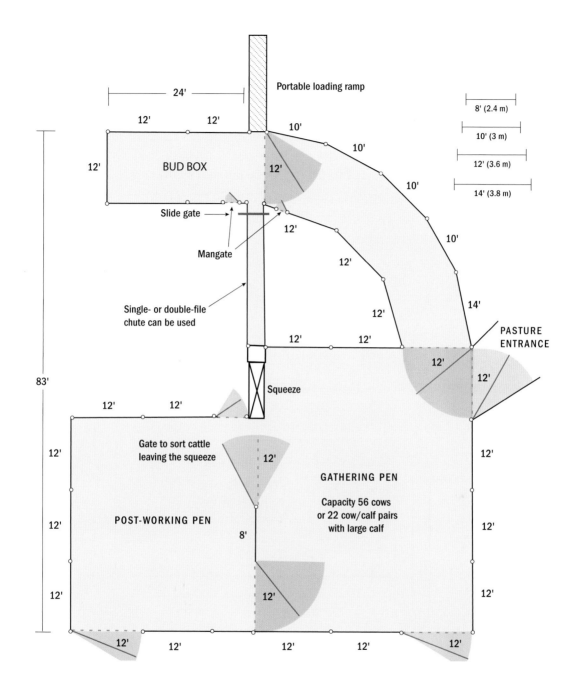

Portable loading ramp

24'

12'   12'

12'

BUD BOX

10'

10'

8' (2.4 m)

10' (3 m)

12' (3.6 m)

14' (3.8 m)

10'

12'

Slide gate

12'

Mangate

12'

12'

Single- or double-file
chute can be used

12'

10'

12'

12'

14'

PASTURE
ENTRANCE

83'

12'   12'

12'

12'

Squeeze

Gate to sort cattle
leaving the squeeze

12'

12'

12'

12'

POST-WORKING PEN

8'

GATHERING PEN

Capacity 56 cows
or 22 cow/calf pairs
with large calf

12'

12'

12'

12'

12'   12'

12'   12'

12'

12'

Bud Box Layout for 50 Cows or 25 Cow/Calf Pairs

This layout can be easily constructed from portable panels and commercially available chute components. The capacity can be easily reduced by making the gathering pen and the post-working pen smaller. The pasture entrance is located to take advantage of the natural tendency of cattle to go back to where they came from. This design is more skill-dependent and requires a more experienced stockperson. Do not use the Bud Box for sheep. For combination cattle and sheep facilities, use of the tub designs is recommended.

Bud Box Layout That Can Be Easily Moved from Pasture to Pasture

This layout, adapted from a design by Ron Gill of Texas A&M University, requires a high level of stockmanship skill, but it has the advantage of ease of portability between multiple pastures. Straight 12-foot (3.5 m) fence panels are easier to set up and take down than curved sections. The double-file alley is commercially available as a trailer-mounted unit. The fence panels can be easily stacked on the chute trailer. When a Bud Box is used, it is essential to have sufficient space in the single-file chutes to hold *all* the cattle put in the Bud Box. When a trailer-mounted alley is used, the double file provides more chute space. The pasture entrance is located to utilize the natural tendency of cattle to go back to where they came from. This will facilitate filling the Bud Box. On these layouts, to increase capacity, add holding pens instead of enlarging the holding pen that is used to fill the Bud Box.

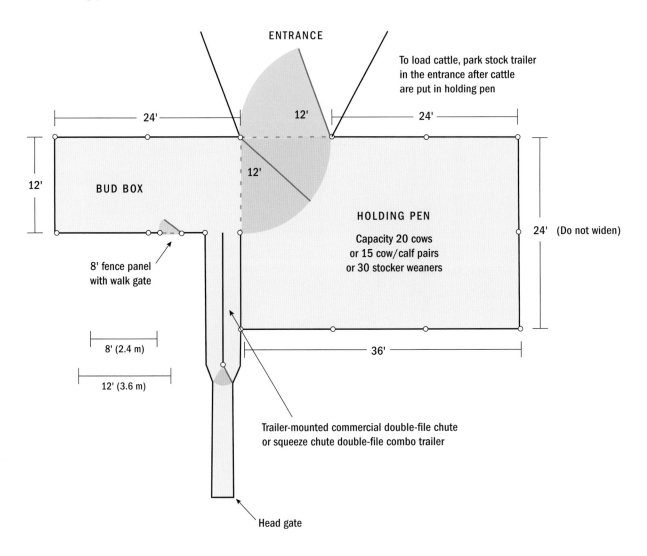

ENTRANCE

To load cattle, park stock trailer in the entrance after cattle are put in holding pen

24'   12'   24'

12'

12'

BUD BOX

8' fence panel with walk gate

HOLDING PEN

Capacity 20 cows
or 15 cow/calf pairs
or 30 stocker weaners

24'   (Do not widen)

8' (2.4 m)

12' (3.6 m)

36'

Trailer-mounted commercial double-file chute
or squeeze chute double-file combo trailer

Head gate

Tub Pivot Design

A single handler works on the ground and takes advantage of the natural tendency of cattle to circle around him or her. The crowd gate is closed to the position shown, and the handler works from the gate pivot point. The only time the crowd gate is closed farther is when one or two remaining animals in the pen become agitated. The outer perimeter fences are solid. Even though this layout can be easily constructed from portable components, it is recommended for either permanent or semipermanent use. It can be easily expanded for larger herds.

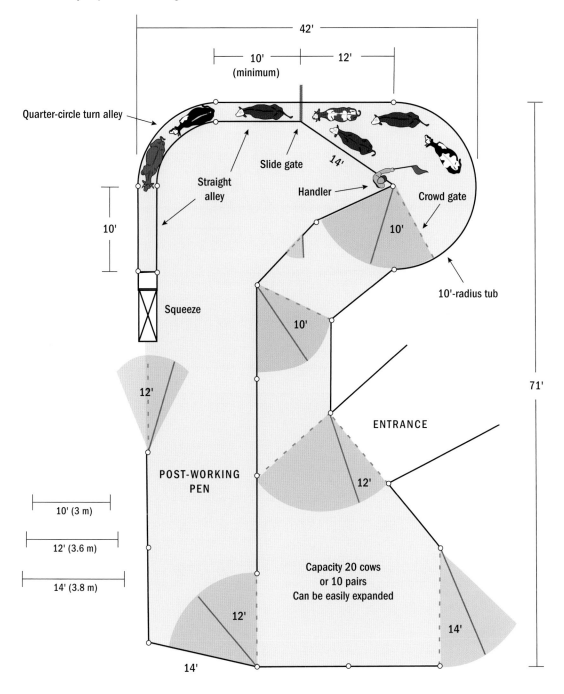

Quarter-circle turn alley

42'

10' (minimum)

12'

Slide gate

14'

Handler

Crowd gate

Straight alley

10'

10'-radius tub

Squeeze

10'

10'

POST-WORKING PEN

12'

ENTRANCE

71'

10' (3 m)

12' (3.6 m)

14' (3.8 m)

12'

Capacity 20 cows or 10 pairs
Can be easily expanded

14'

12'

14'

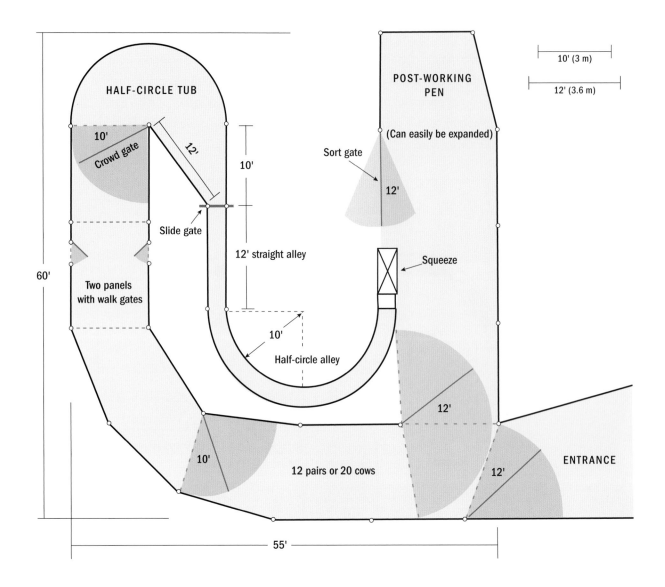

Classic Full Curved Layout

The handler can easily work the tub pivot, but the layout works best with two people. One person works the tub and entrance area, and the other works the single-file chute. Larger herds can be accommodated by expanding the post-working area and creating a gathering pen at the pasture entrance.

All the layouts with a round tub can be easily expanded to handle more cattle. The tub and single- or double-file alley stay the same and the gathering and post-working pens are expanded. The easiest way to do this is to extend the length of the pens.

Sheep and Goat Handling System for a Small Flock

This system can be easily set up with portable fence panels, tub, and alley components. The half-circle design takes advantage of the natural tendency for the animals to go back to where they came from. For sorting sheep, a gate is provided in the single-file chute. Sheep cannot be sorted like cattle. The sort gate in the single-file chute is recommended. See the diagram on page 158 for a sheep-sorting gate.

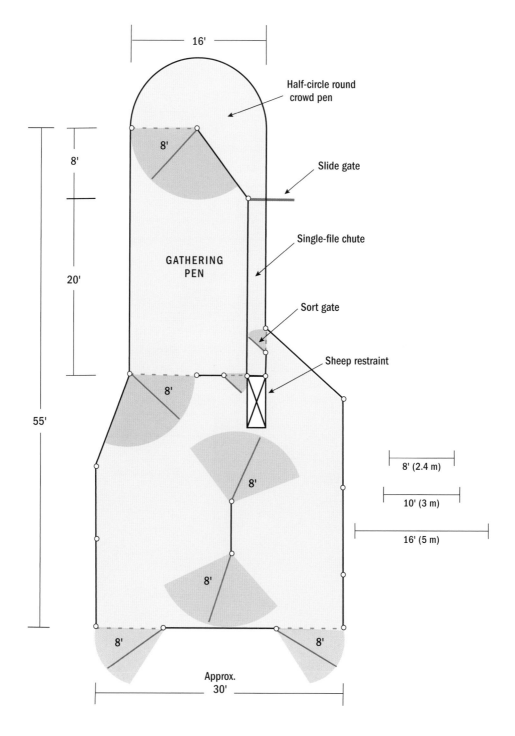

Sheep and Goat Handling System with Option for Expansion

This system can be expanded from one for use with 30 to 40 animals to one for use with 70 to 100 animals. The design uses the principle of the animal's natural behavior to go back to where he came from. The sorting gate is located at the end of the single-file chute to easily sort the sheep into two pens. This design is adapted from the *Sheep Production Handbook* (American Sheep Industry Association, 2015 edition, vol. 8).

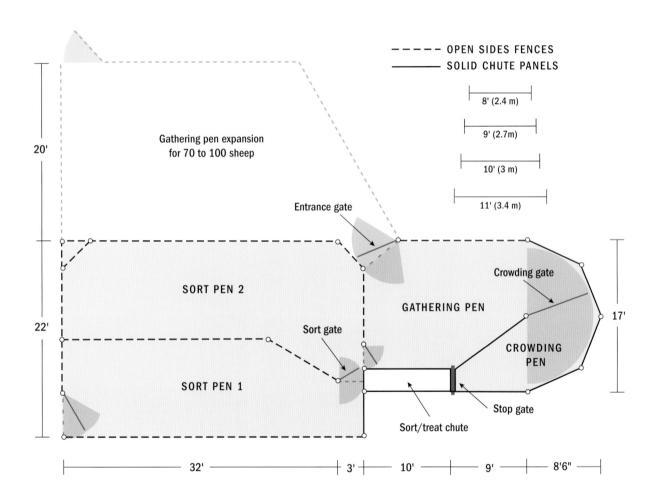

Simple Handling Facility That Can Be Built in a Barn Alley

This design can be used for cattle, sheep, or goats. It is recommended for tame animals that are accustomed to close contact with people. When the gates are closed back against the wall, the animals can easily walk through the alley. If it is located in an alley 10 feet (3 m) wide, a 12-foot (3.5 m) gate is used to form a funnel crowd pen into the single-file chute. On an alley 12 feet (3.5 m) wide, a 14-foot (4.2 m) gate is used. This layout is especially useful for animals housed indoors, and it requires very little space.

CATTLE, SHEEP, OR GOAT CHUTE IN AN EXISTING ALLEY

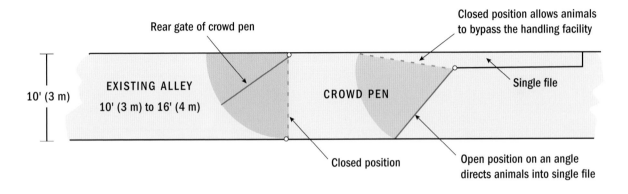

Rear gate of crowd pen

Closed position allows animals to bypass the handling facility

10' (3 m)

EXISTING ALLEY
10' (3 m) to 16' (4 m)

CROWD PEN

Single file

Closed position

Open position on an angle directs animals into single file

7 | CARING FOR ANIMALS AT THE END

How can slaughtering an animal be humane? Doesn't an animal know that it will die at a meat plant?

Early in my career, I spent a lot of time observing animals to answer these questions. On the same day, I'd watch one herd of cattle pass through the squeeze chute for vaccinations at a large feedlot and another go up the chute at the slaughter plant. I discovered that both groups of animals behaved the same way. If they knew they were going to die at the slaughter plant, I reasoned, their behavior there would have been more agitated than it was when they entered the squeeze chute. But blood levels of cortisol, the stress hormone, were similar when taken at the slaughter plant and the vaccinating chute.

Livestock of all species are more frightened of the distractions discussed previously in this book than they are of slaughter. When the distractions are removed, they will usually walk easily up the slaughter chute. I have taken many people on tours of well-managed slaughter plants. They are relieved to discover that the animals remained calm, and that death was quick and painless.

PRINCIPLES

Stress at a well-managed slaughter plant is similar to on-farm handling.

Calm and careful handling during the last few minutes before stunning has multiple benefits.

Correctly stunned animals that are unconscious often have kicking movements.

Raising livestock and poultry is ethical when done right, but the animals must have a life worth living.

Preparing for Stress

At the large commercial plants and feedlots where I observed both slaughter and vaccination, unfamiliar people were handling the cattle in a new place. This caused some fear stress in both locations due to the novelty of the environment. All species of grazing animals get frightened when they are suddenly confronted with things that are novel. Animals that have been raised for their entire lives on the same farm by only one or two people may have an explosive fear reaction when they are transported to a new place such as a fairground, an auction, or a slaughter plant.

Highly agitated behavior can occur in all new places, but it is most likely to happen when the people an animal trusts are absent. One way to prevent this problem is to accustom your animals to different vehicles, people, sights, and sounds before they leave home. This is especially important for animals with more nervous genetics. Knowing that animals are calmer when around someone they trust, some producers walk their own animals up the slaughter chute.

CAREFUL HANDLING IMPROVES MEAT QUALITY

Reducing stress during handling and transport of animals improves both meat quality and animal welfare. Animals bumping into sharp objects can bruise meat. Although the hide may appear undamaged, animals can have severe bruises. Bruises can occur up until the animal is bled. Bruised meat has to be cut out and thrown away. Even old, healed bruises may cause areas of tough meat.

Keeping animals calm the last few minutes before slaughter is essential to prevent tough meat in cattle and pale, soft, exudative (PSE) meat in pigs. PSE pork loses its ability to hold moisture. Agitation, jamming in the chute, or electric prod use shortly before slaughter in the chute that leads to the stunner will increase PSE meat. It is essential to rest pigs for a minimum of one hour before slaughter; the ideal is two or three hours of undisturbed rest time at the plant. Avoid mixing strange pigs in the same pen, because they are likely to fight.

Another condition that occurs in both cattle and sheep is called "dark cutting." This meat is dark and dry, and goes bad more quickly. Dark cutting meat is caused by longer-term stresses such as severe weather a few days before slaughter or animals fighting within a few days of slaughter. Long-term stress depletes the muscle energy (glycogen). To allow the animals' muscles to recover and prevent dark cutting meat, 7 to 10 days after the animals have stopped fighting is required. Strange animals that have not been raised together should be put in separate pens.

Some small-scale producers prefer to avoid the stress of transport and a new environment by slaughtering their animals at home. While it is calmly eating its favorite food, an animal is shot in the head with a gun — an immediate, pain-free method. The carcass is taken promptly to a local plant for processing.

Be sure to find out if killing animals on the farm or ranch is legal in your

country, state, or province. Check out state or provincial meat inspection laws by contracting your country's department of agriculture. Other excellent sources of information are county Cooperative Extension Service agents and area livestock associations.

MAKING TRANSPORT LESS STRESSFUL

There is a great need for more local facilities for harvesting animals. Long transport time to distant facilities may be stressful. Stress during transport may be reduced if the animal is habituated to transport several weeks *before* it has to be transported to the plant. If you own your own stock trailer, feed your animals in it and allow them to explore it. This will make the trailer feel familiar, and they are less likely to become frightened.

Two research studies have shown that livestock were more stressed on their first trip than on their fifth, sixth, or seventh trip. It is also recommended to get your animals accustomed to being around strange people. The animals that become the most stressed when taken to a new place are the ones that seldom see a variety of people and vehicles.

How Do Slaughter Plants Work?

In the United States and many other countries, commercial slaughter plants are required to use methods that will render animals insensible to pain before slaughter procedures begin. In small plants, the most common methods are (1) stunning an animal by shooting it with a pistol, rifle, or captive bolt gun and (2) rendering an animal senseless with an electrical stunner. The next step for each of these methods is cutting an animal's throat to bleed it.

CAPTIVE BOLT STUNNING METHODS

A captive bolt destroys the animal's brain by instantly driving a steel rod into the forehead. It has the same effect as a gun with a bullet, but it is much safer because the method eliminates the risk of a wild bullet hitting something or someone other than the intended target. The bolt is propelled by either compressed air or a blank cartridge. After an animal is shot, the rod is retracted and reset for the next animal.

The gun is placed on the middle of the forehead, not between the eyes. If the gun is aimed between the eyes, the shot will bypass the brain and may fail to kill the animal. To facilitate correct placement of the gun on the animal's forehead, all handling facilities must have nonslip flooring. Slipping causes animals to become fearful and agitated, making it more likely that an animal will move during the process. After an animal is shot, it is normal for the legs to kick. This is a reflex. Even a decapitated animal will kick.

A captive bolt gun is very effective if well maintained. The cartridges must be stored in a dry place. Care and cleaning of the captive bolt gun are absolutely essential to ensure instantaneous insensibility.

FIREARMS FOR STUNNING

For on-farm slaughter, many people prefer to use a firearm. A rifle is strongly recommended because it will propel the bullet at a greater velocity than a pistol. The absolute minimum-size firearm is a .22 long rifle. A larger caliber will be more effective. Large pigs, mature bovine, bulls and bison will require a larger firearm than steers or cows. Large pigs have very heavy skulls.

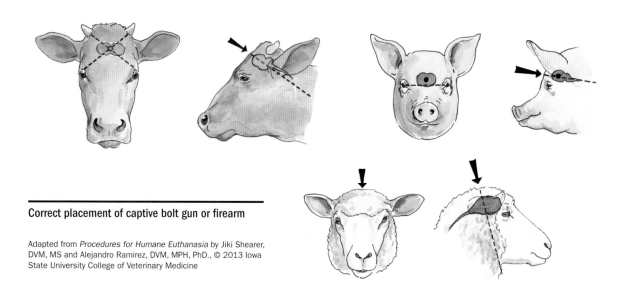

Correct placement of captive bolt gun or firearm

Adapted from *Procedures for Humane Euthanasia* by Jiki Shearer, DVM, MS and Alejandro Ramirez, DVM, MPH, PhD., © 2013 Iowa State University College of Veterinary Medicine

ELECTRICAL STUNNING

When applied correctly, electrical stunning will induce instantaneous insensibility by causing a grand mal epileptic seizure. The electrodes *must* be positioned so that the current passes through the brain. There are two types of electrical stunning: head-only and cardiac arrest.

The head-only method is very common in small plants that do halal (Muslim) slaughter. Head-only stunning does not kill the animal. It produces insensibility for about 30 seconds. When head-only stunning is used, the animal *must* be bled within 15 seconds to prevent it from waking up and regaining consciousness.

In small plants after the electrodes are applied to the head, the electrodes are then reapplied to the chest to induce cardiac arrest. This will prevent return to consciousness. The head stun *must* always be done first to make the animal unconscious. In large plants, a specially designed electrode wand simultaneously passes the current through both the brain and the heart.

ASCERTAINING A SLAUGHTERED ANIMAL'S INSENSIBILITY

Animal slaughter may be disturbing to watch because the animal does not lie still after it is killed. The legs may be kicking and moving. This is a reflex. It is completely *normal* for a *correctly* stunned pig or lamb to kick after electrical stunning. Even when the head is removed, the animal will still kick.

To determine if the animal is unconscious and insensible, you have to ignore leg movement. The head is the most important part of the animal to observe. Spontaneous eye blinking and vocalization must be absent after stunning. There should be no rhythmic breathing evident at the animal's sides. The animal *must* be completely insensible and fully bled before beginning dressing procedures such as skinning and leg removal.

RELIGIOUS SLAUGHTER AND ANIMAL WELFARE

The United States and many other countries have exemptions in their humane slaughter regulations to allow religious

---

More Resources

For further information, refer to the North American Meat Institute (NAMI) Foundation's *Recommended Animal Handling Guidelines & Audit Guide* (available at AnimalHandling.org); my book *Improving Animal Welfare: A Practical Approach*, 2nd ed. (CABI, 2015), on which I served as editor; the Humane Slaughter Association (HSA) in England, the American Veterinary Medical Association (AVMA), or my website (Grandin.com). The AVMA guidelines are titled *AVMA Guidelines for Humane Slaughter of Animals: 2016 Edition*. The NAMI, HSA, and AVMA guidelines can be downloaded free from the Internet. All of the sources mentioned above contain diagrams that show the correct positions for using captive bolt stunners or firearms. Electrode positions for electric stunning are also included. Humane slaughter regulations for the United States and other countries can be easily accessed through search engines on the Internet.

slaughter of animals by throat cutting without stunning for people of the Jewish or Muslim faith. Slaughter without stunning is controversial from an animal welfare standpoint. Research shows mixed results on the question of pain during slaughter without stunning. Many Muslim religious authorities will accept either head-only electric stunning or captive bolt stunning because the heart continues to beat after the stun. They define death as a stopped heart, and the throat cut stops the heart. They want death to result from the throat cut.

It is beyond the scope of this book to describe all of the devices, conveyor systems, and holding boxes used to restrain animals during religious and conventional slaughter. However, all of these systems and the lead-up chutes must have non-slip flooring to prevent panic caused by repeated rapid small slips. To further reduce fear and stress, the animal should be held in a comfortable upright position until it is insensible. An animal should never be dragged or hung upside down by a chain attached to its legs before religious slaughter is performed.

The restraint issue is a separate issue from slaughter without stunning. It is my opinion that problems with the restraint methods cause some of the worst problems with pain and stress to the animal when it is slaughtered without stunning. A review of the scientific literature can be found in my article "Animal Welfare and Society Concerns: Finding the Missing Link," in *Meat Science* 98, no. 3 (November 2014), which you can find easily by typing the title into Google. Information on restraint devices can be found in my book

Livestock Handling and Transport, 4th ed. (CABI, 2014); at my website (Grandin. com); and in papers on the reference list (see page 170).

Compared with sheep or goats, cattle have greater welfare problems during slaughter without stunning. There are two reasons for this. First, sheep and goats are smaller and much easier to restrain than are large cattle. It is easy for a person to hold a lamb in a standing position between his or her legs. The person can easily hold up the head for the throat cut.

The second reason is that there are differences in the anatomy between cows and small ruminants such as sheep and goats. In small ruminants, all the blood vessels that supply the brain with blood are located in the front of the neck. Cattle have some additional blood vessels in the back of the neck that are not cut. Therefore, cattle often take twice as long to lose consciousness. When sheep are cut correctly with a sharp knife, they usually collapse and pass out within 10 seconds.

A very sharp knife should be used that is one-and-a-half times longer than the width of the animal's neck. A long knife is required to prevent the knife from making a gouging cut. There is an easy test for determining whether the knife is sufficiently sharp. To conduct this test, hold a single piece of printer paper by one corner. A really sharp knife held in the other hand should be able to slice the paper.

Maintaining High Standards

All systems should be designed and operated according to the handling principles described in this book. If the cattle or pigs vocalize in a restraint device, stun box, or conveyor system, they are distressed. Vocalization scoring (bellows or squeals) is a good method for detecting severe welfare problems in the stunning area. Each animal is scored as either vocal or silent while he is entering the restrainer or stun box and during the entire time he is in it. If more than 5 percent of the cattle or pigs vocalize in the restraint device, there are problems that need to be corrected. Do not use vocalization scoring for sheep, as they do not indicate their stress or discomfort by vocalizing.

People often ask me if it is stressful for an animal to see another animal killed. When an animal is shot with a gun, it falls down but still looks normal. In most situations, the other animals do not understand what has happened, so they remain calm. If a dead animal is cut up in front of another animal, however, the latter animal will become upset. Never dismember an animal or remove its head in front of other animals. The other animals may become highly frightened and agitated if they see the head removed. Pigs seem to have little reaction to seeing another pig bled because the carcass remains intact.

The slaughter plants that have the best animal welfare standards are usually ones that are audited by a major customer. Major customers such as McDonald's Corporation (in the United States) and Tesco Supermarkets (in England) audit large meat plants. Many smaller plants buy local meat through programs that are either audited by companies such as Whole Foods or certified by an animal welfare organization. (Audit forms and the North American Meat Institute Foundation guidelines can be found at AnimalHandling.org.)

Even though government workers inspect slaughter plants, the plants that are audited by meat buyers usually have higher animal welfare standards. Many people assume that big plants are worse than small ones. This is not true. Bad farms and slaughter plants can be found in huge corporate operations *and* in tiny local places. The attitude of the management team is the most important factor in determining how animals are treated in any given operation.

REGULATIONS GOVERNING THE SALE OF MEAT OR MILK

To sell meat to a large grocery retailer, you must process the animal in an establishment that is inspected by the USDA's Food Safety and Inspection Service. Another term for this is a federally inspected slaughter plant.

For meat sold locally, you will need to check the regulations in your state. Often meat can be marketed as custom exempt. The live animal is sold to the consumer and then the consumer pays a small plant to process it. This meat cannot be sold in stores or go across state lines. In many states, small producers can perform slaughter on the farm. The customer comes to the farm, buys a live animal, and then it is slaughtered.

THOUGHTS ON EATING MEAT

Many people concerned about animal welfare are also concerned that raising animals for food hurts the environment. One form of agriculture that is fully sustainable is raising cattle, sheep, and other ruminants on pasture. Correctly employed grazing methods can actually improve the land. Grasslands that are not grazed deteriorate in the regions of the world where rainfall is not sufficient to support forests.

Livestock and forage crops for feeding livestock are an essential part of a sustainable crop rotation systems that will improve the land. Grazing animals are part of the environment and vital for maintaining soil health. To keep the soil healthy, agriculture should move away from monocropping and perform crop rotation. This will also provide the advantage of reducing diseases and insect infestations of essential food crops. Animals are a natural part of this system to provide fertilizer.

SUSTAINABLE AGRICULTURE

The work of Allan Savory at Holistic Management International in New Mexico has greatly influenced my long-range thinking about use of sustainable grazing practices. I have seen pastures where grazing improved the land, and poorly managed pastures where livestock were allowed to overgraze and strip the land.

Savory explains that to improve the pasture, the grazing patterns of the great herds of bison in years past must be duplicated. A large group of animals must graze a small section of the land at a very high stocking density for only a few days. The grass and all the weedy non-grass plants are eaten, then the animals move to the next section. The sections they have just grazed are rested and the grass is allowed to grow back. Providing previously grazed areas adequate rest time is essential for long-term sustainability.

Similar to the vital role farm animals have in organic agriculture, the high stocking density naturally fertilizes the ground with manure. The hoof action of the herd breaks up the ground's hard surface so rain can penetrate. The natural Great Plains in the middle of the United States were maintained by herds of bison until ruthless people destroyed the animals in the 1800s. Allan Savory and progressive ranchers are grazing herds of cattle or bison in a manner that mimics the original bison grazing patterns, and the land is better for it.

THE ETHICS OF EATING MEAT

Half of the cattle in the United States and Canada are handled in facilities I have designed for slaughter plants. Of course I have done lots of thinking about the ethics of my work. One day, I was standing on a long overhead catwalk at a stockyard and chute system I had designed. As I looked out over a sea of cattle below me, I had the following thought: These animals would never have been born if people had not bred them. They would not have known life.

I feel very strongly that all the animals that are raised for food should be raised in systems where they have a decent life. I am very concerned about welfare problems caused by poor stockmanship or neglect. Overselection of genetic lines to enhance a narrow range of production traits without considering the whole animal is one of my biggest concerns. In some animal genetic lines, indiscriminate single-trait selection (such as selecting only for rapid growth) has resulted in lameness, reduced disease resistance, and highly excitable behavior.

Our relationship with meat animals should be symbiotic. **Symbiosis** is a mutually beneficial relationship between two species. There are many symbiotic relationships between species in the natural world. Ants farm aphids to obtain the sweet substance they secrete. In return, the ants feed them. The relationship should be fully symbiotic and beneficial to both species. When caring people are good stewards of both the animals and the land, the relationship is truly symbiotic.

When examining the ethical issues of slaughter, one must not forget that nature can be very harsh. Death in a well-managed, well-designed slaughter plant is much less frightening or painful than death likely will be in the wild. Some predators such as the big cats kill their prey quickly, but other predatory animals often dine on another animal's guts without killing it first. When an animal quietly walks up the chute at the slaughter plant and death is instantaneous, I feel peaceful. Raising animals for food can be done in an ethical manner. Many other good caretakers of their farm animals feel the same way. We owe farm animals a good life and a calm, painless, humane death.

I wrote this book after I had designed one of my biggest projects, which included a new restrainer system for holding cattle when they die. To design a good restrainer system, you have to really care about the animals it will hold. You have to imagine what it would be like if you were an animal entering it, and you have to respect its every breath, even the last.

If we lose respect for the animals, we lose respect for ourselves. It is a sobering experience to be a caring person, yet design a device that will kill large numbers of animals. When I completed this project, I had a feeling of great satisfaction knowing that the animals were to be treated with care right to the end, but still I cried all the way to the airport.

Additional Reading

American Sheep Industry Association. *Sheep Production Handbook,* 7th ed. American Sheep Industry Association, 2003.

Biggs, D., and C. Biggs, "Gatekeepers: Low Stress Tactics for Moving Cattle," *Stockman Grass Farmer,* September 1996, 36–7.

Boissy, A., C. Terlouw, and P. Le Neindre. "Presence of Cues from Stressed Conspecifics Increases Reactivity to Aversive Events in Cattle: Evidence for the Existence of Alarm Substances in Urine." *Physiology & Behavior* 63, no. 4 (1998): 489–95.

Burns, G. *How Dogs Love Us,* New Harvest Press, New York (2015)

Cooke, R.F., J.D. Arthington, D.B. Araujo, and G.C. Lamb. "Effects of Acclimation to Human Interaction on Performance, Temperament, Physiological Responses, and Pregnancy Rates in Brahman Crossbred Cows." *Journal of Animal Science* 87 (2009): 4125–32.

Cooke, R.F., D.W. Bohnert, B.I. Cappellozza, C.J. Mueller, and T. Del Curto. "Effects of Temperament and Acclimation to Handling on Reproductive Performance of Bos Taurus Beef Females." *Journal of Animal Science* 90 (2012): 3547–55.

Cote, Steve. *Stockmanship: A Powerful Tool for Grazing Lands Management.* USDA Natural Resources Conservation Service and Butte Soil and Water Conservation District, 2004.

Curley, K.O., Jr., J.C. Paschal, T.H. Welsh, Jr., and R.D. Randel. "Technical Note: Exit Velocity as a Measure of Cattle Temperament Is Repeatable and Associated with Serum Concentration of Cortisol in Brahman Bulls." *Journal of Animal Science* 84 (2006): 3100–03.

Edwards, L.N., T.E. Engle, J.A. Correa, M.A. Paradis, T. Grandin, and D.B. Anderson. "The Relationship Between Exsanguination Blood Lactate Concentration and Carcass Quality in Slaughter." *Meat Science* 85, no. 3 (2010): 435–440.

Edwards, L.N., T. Grandin, T.E. Engle, S.P. Porter, M.J. Ritter, A.A. Sosnicki, and D.B. Anderson. "Use of Exsanguination Blood Lactate to Assess the Quality of Pre-Slaughter Pig Handling." *Meat Science* 86, no. 2 (2010): 384–90.

Flörcke, C., T.E. Engle, T. Grandin, and M.J. Deesing. "Individual Differences in Calf Defense Patterns in Red Angus Beef Cows." *Applied Animal Behavior Science* 139 (2012): 203–8.

Fraser, M.D., J.M. Moorby, J.E. Vale, and D.M. Evans. "Mixed Grazing Systems: Benefit Both Upland Biodiversity and Livestock Production." *PLOS ONE* 9, no. 2 (2014): 1–8.

Fulkerson, W.J., and P.A. Jamieson. "Pattern of Cortisol Release in Sheep Following Administration of Synthetic ACTH or Imposition of Various Stressor Agents." *Australian Journal of Biological Science* 35, no. 2 (1982): 215–22.

Graham, A.L., A.D. Hayward, K.A. Watt, J.G. Pilkington, J.M. Pemberton, and D.H. Nussey. "Fitness Correlates of Heritable Variation in Antibody Responsiveness in a Wild Mammal." *Science* 330 (2010): 662–5.

Grandin, T. "Animal Welfare and Society Concerns: Finding the Missing Link." *Meat Science* 98, no. 3 (2014): 461–9.

Grandin, T., "Assessment of Stress During Handling and Transport." *Journal of Animal Science* 75, no. 1 (1997): 249–57.

Grandin, T., "Behavioral Agitation During Handling of Cattle Is Persistent Over Time." *Applied Animal Behaviour Science* 36, no. 1 (1993): 1–9.

Grandin, T. "Euthanasia and Slaughter of Livestock." *Journal of the American Veterinary Medical Association* 204, no. 9 (1994): 1354–60.

Grandin, T. "Factors That Impede Animal Movement in Slaughter Plants." *Journal of the American Veterinary Medical Association* 209 (1996): 757–9.

Grandin, T., and C. Johnson. *Animals in Translation.* Houghton Mifflin Harcourt, 2006.

Grandin, T., and C. Johnson *Animals Make Us Human.* Houghton Mifflin Harcourt, 2009.

Grandin, T., and M.J. Deesing. "Behavioral Genetics and Animal Science." In *Genetics and the Behavior of Domestic Animals,* edited by T. Grandin, 1–30. Academic Press, 1998.

Grandin, T., M.J. Deesing, J.J. Struthers, and A.M. Swinker. "Cattle with Hair Whorl Patterns Above the Eyes Are More Behaviorally Agitated During Restraint." *Applied Animal Behaviour Science* 46, no. 1 (1995): 117–23.

Grandin, T, and M.J. Deesing, ed. *Genetics and the Behavior of Domestic Animals,* 2nd ed. Elsevier Inc., 2014.

Grandin, T., and C. Shively. "How Farm Animals React and Perceive Stressful Situations Such as Handling, Restraint, and Transport." *Animals* 5, no. 4 (2015): 1233–51.

Grandin, T. *Improving Animal Welfare: A Practical Approach,* 2nd Edition, CABI Publishing, Wallingford, Oxfordshire, UK (2015)

Grandin, T., *Livestock Handling and Transport,* 4th ed. CAB International, 2014.

Grandin, T. "Maintenance of Good Animal Welfare Standards in Beef Slaughter Plants by Use of Auditing Programs." *Journal of the American Veterinary Medical Association* 226, no. 3 (2005): 370–3.

Grandin, T. "Observations of Cattle Restraint Devices for Stunning and Slaughtering." *Animal Welfare* 1, no. 2 (1992): 85–91.

Grandin, T. *Recommended Animal Handling Guidelines & Audit Guide,* rev. ed. American Meat Institute Foundation, 2013.

Grandin, T. *Thinking in Pictures,* 2nd ed. Vintage, 2006.

Grandin, T. "Voluntary Acceptance of Restraint by Sheep." *Applied Animal Behaviour Science* 23, no. 3 (1989): 257–61.

Haigh, J.C., and R.J. Hudson, *Farming Wapiti and Red Deer.* Mosby, 1993.

Hargreaves, A.L., and G.D. Hutson. "The Stress Response in Sheep During Routine Handling Procedures." *Applied Animal Behaviour Science* 26, no. 1 (1990): 83–90.

Heffner, R.S., and H.E. Heffner. "Hearing in Large Mammals: Horse (*Equus aballus*) and Cattle (*Bos Taurus*)." *Behavioral Neuroscience* 97, no. 2 (1983): 299–309.

Hemsworth, P.H. "Behavioural Principles of Pig Handling." In *Livestock Handling and Transport,* 3rd ed., edited by T. Grandin, 214–27. CAB International, 2007.

Hutson, G.D. "Behavioral principles of sheep handling." In: T. Grandin (editor) *Livestock Handling and Transport*, 4th ed. CABI Publishing, (2014) pp. 193–217.

Hutson, G.D. "The Effect of Previous Experience on Sheep Movement Through Yards." *Applied Animal Ethology* 6, no. 3 (1980): 233–40.

Hutson, G.D. "The Influence of Barley Food Rewards on Sheep Movement Through a Handling System." *Applied Animal Behaviour Science* 14, no. 3 (1985): 263–73.

Kilgour, R., and H. de Langen. "Stress in Sheep Resulting from Management Practices." *Proceedings of the New Zealand Society of Animal Production* 30 (1970): 65–76.

Lanier, J.L., and T. Grandin. "The Relationship Between Bos Taurus Feedlot Cattle Temperament and Cannon Bone Measurements." *Proceedings, Western Section, American Society of Animal Science* 53 (2002): 97–8.

Lanier, J.L., T. Grandin, R.D. Green, D. Avery, and K. McGee. "The Relationship Between Reaction to Sudden Intermittent Movements and Sounds and Temperament." *Journal of Animal Science* 78, no. 6 (2000): 1467–74.

LeDoux, Joseph E. *The Emotional Brain.* Phoenix, 2004.

Leiner, L., and M. Fendt. "Behavioural Fear and Heartrate Responses of Horses After Exposure to Novel Objects: Effects of Habituation." *Applied Animal Behaviour Science* 131 (2011): 104–9.

Lemaire, G., A. Franzluebbers, P.c. de Faccio Carvalho, and B. Dedieu. "Integrated Crop–Livestock Systems: Strategies to Achieve Synergy Between Agricultural Production and Environmental Quality." *Agriculture, Ecosystems and Environment* 190 (2014): 4–8.

Marchant-Forde, J.N., D.C. Lay, Jr., E.A. Pajor, B.T. Richert, and A.P. Schinckel. "The Effects of Ractopamine on Behavior and Physiology of Finishing Pigs." *Journal of Animal Science* 81, no. 2 (2003): 416–22.

Matthews, L.R. "Deer Handling and Transport." In *Livestock Handling and Transport,* 3rd ed., edited by T. Grandin, 271–94. CAB International, 2007.

Meola, M.G., T. Grandin, P. Burns, and M.J. Deesing. "Hair Whorl Patterns on the Bovine Forehead May Be Related to Breeding Soundness Measures." *Theriogenology* 62, no. 3 (2004): 450–7.

Mitchell, K.D., J.M. Stookey, D.L. Laturnas, J.M. Watts, D.B. Haley, and T. Huyde. "The Effects of Blindfolding on Behavior and Heart Rate in Beef Cattle During Restraint." *Applied Animal Behaviour Science* 85, no. 3 (2004): 233–45.

Panksepp, J. *Affective Neuroscience: The Foundations of Human and Animal Emotions.* Oxford University Press, 1998.

Panksepp, J. "The Basic Emotional Circuits of Mammalian Brains: Do Animals Have Emotional Lives?" *Neuroscience & Biobehavioral Reviews* 35 (2011): 1791–1804.

Phillips, M., T. Grandin, W. Graffman, N.A. Irlbeck, and R.C. Cambre. "Crate Conditioning of Bongo (*Tragelephus eurycerus*) for Veterinary and Husbandry Procedures at Denver Zoological Gardens." *Zoo Biology* 17, no. 1 (1998): 25–32.

Pollard, J.C., and P.R. Wilson. "Welfare of Farmed Deer in New Zealand 1. Management Practices." *New Zealand Veterinary Journal* 50, no. 6 (2002): 214–20.

Price, E.O., and S.J.R. Wallach. "Physical Isolation of Hand-Reared Hereford Bulls Increases Their Aggressiveness Toward Humans." *Applied Animal Behaviour Science* 27, no. 4 (1990): 277–85.

Provenza, F.D. *Foraging Behavior: Managing to Survive a World of Change.* Utah State University Press, 2003.

Robins, A., and C. Phillips. "Lateralized Visual Processing in Domestic Cattle Herds Responding to Novel or Familiar Stimuli." *Laterality* 15, no. 5 (2010): 514–34.

Sandem, A.I., A.M. Janczak, R. Salte, and B.O. Braastad. "The Use of Diazepam as a Pharmacological Validation of Eye White as an Indicator of Emotional State in Dairy Cows." *Applied Animal Behaviour Science* 96, no. 3 (2006): 177–83.

Shinozaki, et al. "Topography of ganglion cells and photoreceptors in sheep retina," *Journal of Comparative Neurology* 518 (2010): 2305–2315.

Shivley, C., T. Grandin, and M.J. Deesing. "Behavioral Laterality and Facial Hair Whorls in Horses." *Journal of Equine Veterinary Science* 44 (2016): 62–6.

Simon, G.E., et al. "Assessing cow-calf welfare, Part 2, Risk factors for beef cow health and behavior and stockperson handling," *Journal of Animal Science* 94 (2016): 3488–3500.

Smith, Burt. *Moving 'Em: A Guide to Low Stress Animal Handling. Graziers Hui,* 1998.

Stockman, C.A., T. Collins, A.L. Barnes, D. Miller, S.L. Wickham, D.T. Beatty, D. Blache, F. Wemelsfelder, and P.A. Fleming. "Qualitative Behaviorial Assessment and Physiological Measurement of Cattle Naïve and Habituated to Road Transport." *Animal Production Science* 51, no. 3 (2010): 240–9.

Talling, J.C., N.K. Waran, C.M. Wathes, and J.A. Lines. "Sound Avoidance by Domestic Pigs Depends Upon Characteristics of the Signal." *Applied Animal Behavior Science* 58, no. 3 (1998): 255–66.

Tanida, H., A. Miura, T. Tanaka, and T. Yoshimoto. "Behavioral Responses of Piglets to Darkness and Shadows." *Applied Animal Behaviour Science* 49, no. 2 (1996): 173–83.

Vetters, M.D.D., et al. "Comparison of flight speed and exit score as measurement of temperament in beef cattle," *Journal of Animal Science* 93 (2013): 374–381.

Vogel, K.D., G. Badtram, J.R. Claus, T. Grandin, S. Turpin, R.E. Weyker, and E. Voogd. "Head-only Followed by Cardiac Arrest Electrical Stunning Is an Effective Alternative to Head-only Electrical Stunning in Pigs." *Journal of Animal Science* 89, no. 5 (2010): 1412–8.

Voisinet, B.D., T. Grandin, J.D. Tatum, S.F. O'Connor, and J.J. Struthers. "Feedlot Cattle with Calm Temperaments Have Higher Average Daily Gains Than Cattle with Excitable Temperament." *Journal of Animal Science* 75, no. 4 (1997): 892–6.

Warner, R.D., D.M. Ferguson, J.J. Cottrell, and B.W. Knee. "Acute Stress Induced by the Preslaughter Use of Electrical Prodders Causes Tougher Beef Meat." *Australian Journal of Experimental Agriculture* 47 (2007): 782–8.

Waynert, D.F., J.M. Stookey, K.S. Schwartzkopf-Genswein, J.M. Watts, and C.S. Waltz. "The Response of Beef Cattle to Noise During Handling." *Applied Animal Behaviour Science* 62, no. 1 (1999): 27–42.

Woiwode, R., et al. "Effects of initial handling practices on behavior and average daily gain of fed steers," *International Journal of Livestock Production* 7(3) (2015): 12–18 (open access paper).

Woiwode, R., et al. "Validation of the Beef Quality Assurance Feedyard Assessment of Cattle Handling," *Professional Animal Scientist* 32 (2016): 750–757 (open access).

Web Resources

LIVESTOCK BEHAVIOR AND HANDLING

Animal Behavior Society
animalbehaviorsociety.org
Excellent links and educational materials; publisher of *Journal of Animal Behavior*

American Veterinary Medical Association
800-248-2862
avma.org

Animal Transportation Association
animaltransportationassociation.org

Bud Williams
stockmanship.com
Expert cattle herding training, advice, musings, and practical applications
blm.gov/or/programs/nrst/files/Stockmanship_Book.pdf
Stockmanship – A Powerful Tool for Grazing Land Management by Steve Cote (Open Access)

Center for Food Animal Well-Being Purdue University
ansc.purdue.edu/CAWB
Animal behavior research

Curt Pate Stockmanship
https://curtpatestockmanship.com
Livestock handling demonstrations

Federation of Animal Science Societies
fass.org
Source of the third edition of *The Guide for the Care and Use of Agricultural Animals in Research and Teaching* (2010)
Proper use of livestock driving tools

Grandin Livestock Handling Systems, Inc.
grandinlivestockhandlingsystems.com
Numerous photographs of systems designed by Temple Grandin and Mark Deesing

Humane Slaughter Association
hsa.org.uk

International Society for Applied Ethology
applied-ethology.org
Global organization of scientists studying animal behavior

Lopez Community Land Trust
lopezclt.org
Information about mobile slaughterhouse and meat-processing units that deliver slaughtering services to rural livestock producing communities

ManagingWholes.com
managingwholes.com
Low-stress handling articles and links by Bud Williams and other authors

Manitoba Pork Council
Manitobapork.com
Basics and principles of pig handling and behavior

National Bison Association
bisoncentral.com
Comprehensive information on the bison industry, outlook raising bison, and history; many good links to other resources on bison

North American Meat Institute
animalhandling.org
Printable slaughter plant audit forms; Glass Walls video tours of beef, pork, and lamb slaughter plants.

Prairie Swine Centre
prairieswine.com
Research on pig behavior

Effective Stockmanship
Effectivestockmanship.com
Services and resources for low-stress cattle

Dr. Temple Grandin

www.grandin.com

Articles and research about cattle and other stock animals; video collection of cattle handling; also contains more information on hair whorls and drawings of restraint devices for slaughter

DATABASES

Animal Welfare Information Center USDA National Agricultural Library

awic.nal.usda.gov

Includes a database of animal behavior and welfare information

Association for the Study of Animal Behavior

asab.nottingham.ac.uk

Information on animal behavior

Center for Animal Welfare Science Purdue University

vet.purdue.edu/CAWS/

Links to journals and animal behavior research

Commonwealth Scientific and Industrial Research Organization (CSIRO)

csiro.au

Searchable database of Australian research on agriculture and livestock

Google Scholar

scholar.google.com

Searchable database of scientific literature

PubMed
U.S. National Library of Medicine, National Center for Biotechnology Information

ncbi.nlm.nih.gov/pubmed

Serves as an open access portal to older articles in *Journal of American Science*. Searchable articles in veterinary and medical journals

ResearchGate

Researchgate.net

Many scientists have their papers available on this website; membership is free

ScienceDirect

sciencedirect.com

Search engine for scientific journal articles

LIVESTOCK PUBLICATIONS AND ASSOCIATIONS

American Sheep Industry Assoc.

sheepusa.org

National Sheep Organization

Beef Magazine

beefmagazine.com

Current and archived *Beef* magazine articles and information on cattle production

Cattle Today

cattletoday.com

Current cattle news and USDA livestock market report: articles on cattle breeds, diseases, and many links for cattle ranchers; lists of cattle breed associations

Drovers CattleNetwork

cattlenetwork.com

Beef business source with resources and articles

National Cattlemen's Beef Association
beefusa.org
National organization of cattle producers

National Hog Farmer
Nationalhogfarmer.com
Current and archived articles on pork production

National Pork Board Checkoff
pork.org
Pork quality assurance certification

National Pork Producers Council
Nppc.org
Pork Producers Association

The Stockman Grass Farmer
stockmangrassfarmer.com
Magazine with a focus on grass and forage production for cattle

Western Livestock Journal
wlj.net
General information about the livestock industry

ORGANIC AND NATURAL LIVESTOCK

Certified Organic Associations of British Columbia
certifiedorganic.bc.ca
Lots of information on organic beef and livestock

The Livestock Conservancy
livestockconservancy.org
Organization that seeks to conserve livestock diversity and heritage breeds

National Organic Program Agricultural Marketing Service
www.ams.usda.gov/about-ams/programs-offices/national-organic-program
USDA site with information on organic beef and farming

National Sustainable Agriculture Coalition (NSAC)
Sustainableagriculture.net
Supports sustainable and organic agriculture

National Sustainable Agriculture Information Service – ATTRA National Center for Appropriate Technology
attra.ncat.org
Articles on intensive grazing and organic farming practices and trends

Soil Association
soilassociation.org
United Kingdom website on organic and sustainable farming with information on organic beef

Sustainable Agriculture Research and Education (SARE)
sare.org
Grants and education to advance innovations in sustainable agriculture

Agricultural Sustainability Institute University of California, Davis
Asi.ucdavis.edu
Lots of information on sustainable agriculture programs

Index

Page numbers in *italic* indicate photos and illustrations; page numbers in **bold** indicate charts.

PHOTO CREDITS

More Essential Books from Storey

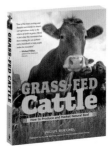
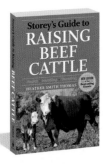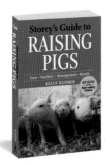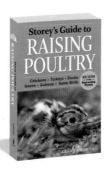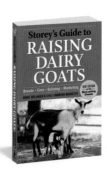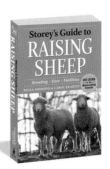